Praise for *Life at the Speed of Light*

"A fascinating glimpse at a scientific frontier."

—*Kirkus Reviews* (starred review)

"Venter shares spellbinding stories from the frontiers of genomics—researchers creating living toolboxes out of mechanisms co-opted from varied life forms . . . Venter instills awe for biology as it is, and as it might become in our hands."

—*Publishers Weekly*

"Readers will marvel at the potential that genetic engineering holds for making food, purifying water, generating energy, and curing diseases . . . and will thank Venter for an insider's perspective on epoch-making science."   —*Booklist*

"One of the world's leading scientists delivers a history of molecular biology and its many ramifications depicted as it has been and will continue to be, creator of the golden age of modern biology. His style is that of a dispatch from the front, urgent and engaged, as only a participant could write it, and the best of its genre since James D. Watson's *The Double Helix*."

—Edward O. Wilson, University Research Professor
Emeritus, Harvard University

"In 1943 Erwin Schrödinger asked the question 'What is life?' Craig Venter decided to find out. The first step in understanding something is to take it apart. The next step is to put it back together. Finally, prove that you haven't neglected anything by building it yourself from scratch. In this articulate, precise, and uniquely first-person report from the front lines between biology and technology—commendably addressing the role of proteins as well as the nucleotides that code for them—Craig Venter sheds new light on Schrödinger's question, while explaining how his own pioneering work reading and writing genetic sequences between living cells and computers is enabling life as we know it to take the first steps toward becoming something else. A landmark account."

—George Dyson, author of *Turing's Cathedral* and
Darwin Among the Machines

PENGUIN BOOKS

LIFE AT THE SPEED OF LIGHT

J. Craig Venter is best known for sequencing the human genome. He is the founder, chairman, and CEO of the J. Craig Venter Institute, a not-for-profit research organization dedicated to genomic research. He is also the founder and CEO of Synthetic Genomics, Inc. He is the recipient of numerous awards and honorary degrees, including the 2008 United States National Medal of Science. He lives in La Jolla, California.

LIFE AT THE
 SPEED OF LIGHT

From the Double Helix to the
Dawn of Digital Life

J. CRAIG VENTER

PENGUIN BOOKS

PENGUIN BOOKS
Published by the Penguin Group
Penguin Group (USA) LLC
375 Hudson Street
New York, New York 10014

USA | Canada | UK | Ireland | Australia | New Zealand | India | South Africa | China
penguin.com
A Penguin Random House Company

First published in the United States of America by Viking Penguin,
a member of Penguin Group (USA) LLC, 2013
Published in Penguin Books 2014

THE LIBRARY OF CONGRESS HAS CATALOGED THE HARDCOVER
EDITION AS FOLLOWS:
Venter, J. Craig.
Life at the speed of light : from the double helix to the dawn of digital life / J. Craig Venter.
pages   cm
Includes bibliographical references and index.
ISBN 978-0-670-02540-4 (hc.)
ISBN 978-0-14-312590-7 (pbk.)
1. Science—Social aspects. 2. Biology—Philosophy. 3. Artificial life. 4. Genomics. I. Title.
Q175.5.V44 2013
303.48'3—dc23
2013017049

Printed in the United States of America
10 9 8 7 6 5 4 3 2

Set in Warnock Pro Light
Designed by Carla Bolte

To the team that contributed to making the first synthetic cell a reality: Mikkel A. Algire, Nina Alperovich, Cynthia Andrews-Pfannkoch, Nacyra Assad-Garcia, Kevin C. Axelrod, Holly Baden-Tillson, Gwynedd A. Benders, Anushka Brownley, Christopher H. Calvey, William Carrera, Ray-Yuan Chuang, Jainli Dai, Evgeniya A. Denisova, Tom Deernick, Mark Ellisman, Nico Enriquez, Robert Friedman, Daniel G. Gibson, John I. Glass, Jessica Hostetler, Clyde A. Hutchison III, Prabha Iyer, Radha Krishnakumar, Carole Lartigue, Matt Lewis, Li Ma, Mahir Maruf, Admasu Melanke, Chuck Merryman, Michael G. Montague, Monzia M. Moodie, Vladimir N. Noskov, Prashanth P. Parmar, Quang Phan, Rembert Pieper, Thomas H. Segall-Shapiro, Hamilton O. Smith, Timothy B. Stockwell, Lijie Sun, Granger Sutton, Yo Suzuki, David W. Thomas, Christopher E. Venter, Sanjay Vashee, Shibu Yooseph, Lei Young, and Jayshree Zaveri.

Contents

LIFE AT THE SPEED OF LIGHT

1 Dublin, 1943–2012

How can the events in space and time, which take place within the boundaries of a living organism, be accounted for by physics and chemistry? . . . The obvious inability of present-day physics and chemistry to account for such events is no reason at all for doubting that they will be accounted for by those sciences.

—Erwin Schrödinger, *What Is Life?* (1944)[1]

"What is life?" Only three simple words, and yet out of them spins a universe of questions that are no less challenging. What precisely is it that separates the animate from the inanimate? What are the basic ingredients of life? Where did life first stir? How did the first organisms evolve? Is there life everywhere? To what extent is life scattered across the cosmos? If other kinds of creatures do exist on exoplanets, are they as intelligent as we are, or even more so?

Today these questions about the nature and origins of life remain the biggest and most hotly debated in all of biology. The entire discipline depends on it, and though we are still groping for all the answers, we have made huge progress in the past decades toward addressing them. In fact, we have advanced this quest further in living memory than during the ten thousand or so generations that modern humans have walked on the planet.[2] We have now entered what I call "the digital age of biology," in which the once distinct domains of computer codes and those that program life are beginning to merge, where new synergies are emerging that will drive evolution in radical directions.

If I had to pick the moment at which I believe that modern biological science was born, it would be in February 1943, in Dublin, when Erwin Schrödinger (1887–1961), an Austrian physicist, focused his mind on the central issue in all of biology. Dublin had become Schrödinger's home in 1939, in part to escape the Nazis, in part because of its tolerance of his unconventional domestic life (he lived in a ménage à trois and pursued "tempestuous sexual adventures" for inspiration[3]), and in part because of the initiative of the then-Taoiseach (Gaelic for prime minister) of Ireland, Éamon de Valera, who had invited him to work there.

Schrödinger had won the Nobel Prize in 1933 for his efforts to devise an equation for quantum waves, one with the power to explain the behavior of subatomic particles, the universe itself, and everything in between. Now, ten years later, speaking under the auspices of the Dublin Institute for Advanced Studies, which he had helped to found with de Valera, Schrödinger gave a series of three lectures in Trinity College, Dublin, that are still quoted today. Entitled "What is Life? The Physical Aspect of the Living Cell," the talks were inspired in part by his father's interest in biology and in part by a 1935 paper[4] that resulted from an earlier encounter between physics and biology in prewar Germany. German physicists Karl Zimmer and Max Delbrück had then worked with the Russian geneticist Nikolai Timoféeff-Ressovsky to develop an estimate of a gene's size ("about 1,000 atoms"), based on the ability of X-rays to damage genes and cause mutations in fruit flies.

Schrödinger began the series at 4:30 P.M. on Friday, February 5, with the Taoiseach sitting before him in the audience. A reporter from *Time* magazine was present and described how "crowds were turned away from a jam-packed scientific lecture. Cabinet ministers, diplomats, scholars and socialites loudly applauded a slight, Vienna-born professor of physics [who] has gone beyond the ambitions of any other mathematician." The next day, *The Irish Times* carried an article on "The Living Cell and the Atom," which began by describing Schrödinger's aim to account for events within a living cell by using chemistry and physics alone. The lecture was so popular that he had to repeat the entire series on the following Mondays.

Schrödinger converted his lectures into a small book that was published the following year, two years before my own birth. *What Is Life?*

has gone on to influence generations of biologists. (Fifty years after he had delivered these remarkable talks, Michael P. Murphy and Luke A. J. O'Neill, of Trinity, celebrated the anniversary by inviting outstanding scientists from a range of disciplines—a prestigious guest list that included Jared Diamond, Stephen Jay Gould, Stuart Kauffman, John Maynard Smith, Roger Penrose, Lewis Wolpert, and the Nobel laureates Christian de Duve and Manfred Eigen—to predict what the next half-century might hold.) I have read *What Is Life?* on at least five different occasions, and each time, depending on the stage of my career, its message has taken on different meanings along with new salience and significance.

The reason that Schrödinger's slim volume has proved so influential is that, at its heart, it is simple: it confronted the central problems of biology—heredity and how organisms harness energy to maintain order—from a bold new perspective. With clarity and concision he argued that life had to obey the laws of physics and, as a corollary, that one could use the laws of physics to make important deductions about the nature of life. Schrödinger observed that chromosomes must contain "some kind of code-script determining the entire pattern of the individual's future development." He deduced that the code-script had to contain "a well-ordered association of atoms, endowed with sufficient resistivity to keep its order permanently" and explained how the number of atoms in an "aperiodic crystal" could carry sufficient information for heredity. He used the term "crystal" to suggest stability, and characterized it as "aperiodic," which unlike a periodic, repeating pattern (which, explained *The Irish Times*, is like "a sheet of ordinary wallpaper when compared with an elaborate tapestry"), could have a high information content. Schrödinger argued that this crystal did not have to be extremely complex to hold a vast number of permutations and could be as basic as a binary code, such as Morse code. To my knowledge, this is the first mention of the fact that the genetic code could be as simple as a binary code.

One of the most remarkable properties of life is this ability to create order: to hone a complex and ordered body from the chemical mayhem of our surroundings. At first sight this capability seems to be a miracle that defies the gloomy second law of thermodynamics, which states that everything tends to slide from order toward disorder. But this law only applies to a "closed system," like a sealed test tube, while living things are open (or are a small part of a larger closed system), being permeable to energy and mass in their surroundings. They expend large amounts of energy to create order and complexity in the form of cells.

Schrödinger dedicated much of his lecture to the thermodynamics of life, a topic that has been relatively underinvestigated compared with his insights into genetics and molecular biology. He described life's "gift of concentrating a 'stream of order' on itself and thus escaping the decay into 'atomic chaos'—and of 'drinking orderliness' from a suitable environment." He had worked out how an "aperiodic solid" had something to do with this creative feat. Within the code-script lay the means to rearrange nearby chemicals so as to harness eddies in the great stream of entropy and to make them live in the form of a cell or body.

Schrödinger's hypothesis would inspire a number of physicists and chemists to turn their attention to biology after they had become disenchanted with the contribution of their fields to the Manhattan Project, the vast effort to build the atomic bomb during the Second World War. At the time of Schrödinger's lecture the scientific world believed that proteins and not DNA formed the basis of the genetic material. In 1944 came the first clear evidence that DNA was in fact the information-carrier, not protein. Schrödinger's book motivated the American James Watson and Briton Francis Crick to seek that code-script, which ultimately led them to DNA and to discover the most beautiful structure in all biology, the double helix, within whose turns lay the secrets of all inheritance. Each strand of the double helix is complementary to the other, and they therefore run in opposite (anti-parallel) directions. As a result the double helix can unzip down the middle, and each side can serve as a pattern or template for the other, so that the DNA's information can be copied and passed to progeny. On August 12, 1953, Crick sent Schrödinger a letter indicating as much, adding that "your term 'aperiodic crystal' is going to be a very apt one."

In the 1960s the details of precisely how this code works were uncovered and then unraveled. This led to the formulation by Crick in 1970 of the "central dogma," which defined the way that genetic information flows through biological systems. In the 1990s I would lead the team to read the first genome of a living cell and then lead one of the two teams that would read the human code-script, in a highly publicized race with Watson and others that was often heated, fractious, and political. By the turn of the millennium, we had our first real view of the remarkable details of the aperiodic crystal that contained the code for human life.

Implicit in Schrödinger's thinking was the notion that this code-script had been sending out its signals since the dawn of all life, some four billion years ago. Expanding upon this idea, biologist and writer Richard Dawkins came up with the evocative image of a river out of Eden.[5] This slow-flowing river consists of information, of the codes for building living things. The fidelity of copying DNA is not perfect, and together with oxidative and ultraviolet damage that has taken place in the course of generations, enough DNA changes have occurred to introduce new species variations. As a result, the river splits and bifurcates, giving rise to countless new species over the course of billions of years.

Half a century ago the great evolutionary geneticist Motoo Kimura estimated that the amount of genetic information has increased by one hundred million bits over the past five hundred million years.[6] The DNA code-script has come to dominate biological science, so much so that biology in the twenty-first century has become an information science. Sydney Brenner, the Nobel Prize–winning South African biologist, remarked that the code-script "must form the kernel of biological theory."[7] Taxonomists now use DNA bar codes to help distinguish one species from another.[8] Others have started to use DNA in computation,[9] or as a means to store information.[10] I have led efforts not only to read the digital code of life but also to write it, to simulate it within a computer, and even to rewrite it to form new living cells.

On July 12, 2012, almost seven decades after Schrödinger's original lectures, I found myself in Dublin, at the invitation of Trinity College. I was asked to return to Schrödinger's great theme and attempt to provide new insights and answers to the profound question of defining life, based on modern science. Everyone is still interested in the answer, for obvious reasons, and I have very personal ones, too. As a young corpsman in Vietnam, I had learned to my amazement that the difference between the animate and inanimate can be subtle: a tiny piece of tissue can distinguish a living, breathing person from a corpse; even with good medical care, survival could depend in part on the patient's positive thinking, on remaining upbeat and optimistic, proving a higher complexity can derive from combinations of living cells.

At 7:30 on a Thursday evening, with the benefit of decades of progress in molecular biology, I walked up to the same stage on which Schrödinger appeared, and like him appearing before the Taoiseach, in what was now the Examination Hall of Trinity College, a matchless backdrop. Under a vast chandelier, and before portraits of the likes of William Molyneux and Jonathan Swift, I gazed into an audience of four hundred upturned faces and the bright lights of cameras of every kind and description. Unlike Schrödinger's lectures, I knew my own would be recorded, live-streamed, blogged, and tweeted about as I once again tackled the question that my predecessor had done so much to answer.

Over the next sixty minutes I explained how life ultimately consists of DNA-driven biological machines. All living cells run on DNA software, which directs hundreds to thousands of protein robots. We have been digitizing life for decades, since we first figured out how to read the software of life by sequencing DNA. Now we can go in the other direction by starting with computerized digital code, designing a new form of life, chemically synthesizing its DNA, and then booting it up to produce the actual organism. And because the information is now digital we can send it anywhere at the speed of light and re-create the DNA and life at the other end. Sitting next to Taoiseach Enda Kenny was my old self-proclaimed rival, James Watson. After I had finished, he climbed onto the stage, shook my hand, and graciously congratulated me on "a very beautiful lecture."[11]

Life at the Speed of Light, which is based in part on my Trinity College lecture, is intended to describe the incredible progress that we have made. In the span of a single lifetime, we have advanced from Schrödinger's "aperiodic crystal" to an understanding of the genetic code to the proof, through construction of a synthetic chromosome and hence a synthetic cell, that DNA is the software of life. This endeavor builds on tremendous advances over the last half-century, made by a range of extraordinarily gifted individuals in laboratories throughout the world. I will provide an overview of these developments in molecular and synthetic biology, in part to pay tribute to this epic enterprise, in part to acknowledge the contributions made by key leading scientists. My aim is not to offer a comprehensive history of synthetic biology but to shed a little light on the power of that extraordinarily cooperative venture we call science.

DNA, as digitized information, is not only accumulating in computer databases but can now be transmitted as an electromagnetic wave at or near the speed of light, via a biological teleporter, to recreate proteins, viruses, and living cells at a remote location, perhaps changing forever how we view life. With this new understanding of life, and the recent advances in our ability to manipulate it, the door cracks open to reveal exciting new possibilities. As the Industrial Age is drawing to a close, we are witnessing the dawn of an era of biological design. Humankind is about to enter a new phase of evolution.

2 Chemical Synthesis as Proof

This type of synthetic biology, a grand challenge to create artificial life, also challenges our definition-theory of life. If life is nothing more than a self-sustaining chemical system capable of Darwinian evolution and if we truly understand how chemistry might support evolution, then we should be able to synthesize an artificial chemical system capable of Darwinian evolution. If we succeed, the theories that supported our success will be shown to be empowering. . . . In contrast, if we fail to get an artificial life form after an effort to create a chemical system . . . , we must conclude that our theory of life is missing something.

—Steven A. Benner, 2009[1]

Humans have long been fascinated with the notion of artificial life. From the medieval homunculus of Paracelsus and the golem of Jewish folklore to the creature of Mary Shelley's *Frankenstein* and the "replicants" of *Blade Runner*, mythology, legend, and popular culture are replete with tales of synthetic and robotic life. However, devising a precise definition that captures the distinction between life and non-life, or between biological life and machine life, has been a major and continuing challenge for science and philosophy alike.

For centuries, a principal goal of science has been, first, to understand life at its most basic level and, second, to learn to control it. The German-born American biologist Jacques Loeb (1859–1924) was perhaps the first true biological engineer. In his laboratories in Chicago, New York, and Woods Hole, Massachusetts, he constructed what he referred to as "durable machines" in his 1906 book, *The Dynamics of Living Matter.*[2] Loeb made two-headed worms and, most famously, caused the eggs of sea urchins to begin embryonic development without being fertilized by sperm.[3] No wonder Loeb became the inspiration for the character of Max Gottlieb in Sinclair Lewis's Pulitzer Prize–winning novel *Arrowsmith*, published in 1925, the first major work of fiction to idealize pure science, including the antibacterial power of viruses called bacteriophages.

Philip J. Pauly's *Controlling Life: Jacques Loeb and the Engineering Ideal in Biology* (1987) cites a letter sent in 1890 from Loeb to the Viennese physicist and philosopher Ernst Mach (1838–1916), in which Loeb stated, "The idea is now hovering before me that man himself can act as a creator, even in living Nature, forming it eventually according to his will. Man can at least succeed in a technology of living substance [einer Technik der lebenden Wesen]." Fifteen years later Loeb prefaced a volume of his scientific papers with the explanation that "in spite of the diversity of topics, a single leading idea permeates all the papers of this collection, namely, that it is possible to get the life-phenomena under our control, and that such a control and nothing else is the aim of biology."

The origins of Loeb's mechanistic view of life can in fact be glimpsed centuries before his correspondence with Mach. Some of the earliest theories of life were "materialistic" in contrast to those that relied on a nonphysical process that lay outside material nature and relied on a supernatural means of creation. Empedocles (c. 490–430 B.C.) argued that everything—including life—is made up of a combination of four eternal "elements" or "roots of all": earth, water, air, and fire. Aristotle (384–322 B.C.), one of the original "materialists," divided the world into the three major groups of animal, vegetable, and mineral, a classification that is still taught in schools today. In 1996 my team sequenced the first Archaeal genome. This sequence was touted by many as proof that the Archaea, as first proposed by American microbiologist Carl Woese, represents a third branch of life. When the news broke, the television anchor Tom Brokaw asked rhetorically, "We have animal, vegetable, and mineral. What could the new branch be?"

As understanding deepened, thinkers became more ambitious. Among the Greeks, the idea of altering nature to suit human desires or seeking to control it was seen as absurd. But since the birth of the Scientific Revolution, in the sixteenth century, a principal goal of science has not only been to investigate the cosmos at its most basic level but also to master it. Francis Bacon (1561–1626), the English polymath who gave us empiricism, in effect remarked that it was better to show than merely to tell: the Greeks "assuredly have that which is characteristic of boys; they are prompt to prattle but cannot generate; for their wisdom abounds in words but is barren of works . . . From all these systems of the Greeks, and their ramifications through particular sciences, there can hardly after the lapse of so many years be adduced a single experiment which tends to relieve and benefit the condition of man."

In Bacon's utopian novel, *New Atlantis* (1623),[4] he outlined his vision of a future marked by human discovery and even envisaged a state-sponsored scientific institution, Salomon's House,[5] in which the goal is to "establish dominion over Nature and effect all things possible." His novel describes experiments with "beasts and birds" and what sounds like genetic modification: "By art likewise we make them greater or smaller than their kind is, and contrariwise dwarf them and stay their growth; we make them more fruitful and bearing than their kind is, and contrariwise barren and not generative. Also we make them differ in color, shape, activity, many ways." Bacon even alludes to the ability to design life: "Neither do we this by chance, but we know beforehand of what matter and commixture, what kind of those creatures will arise."[6]

In this search for power over Nature, science sees a union of the quest for understanding with the service of man. René Descartes (1596–1650), a pioneer of optics whom we all associate with "I think, therefore I am," also looked forward in his *Discourse on the Method* (1637) to a day when mankind would become "masters and possessors of nature." Descartes and his successors extended mechanistic explanations of natural phenomena to biological systems and then explored its implications. From the very birth of this great endeavor, however, critics have expressed concerns that wider moral and philosophical issues were being neglected in the quest for efficient mastery over nature. With the Faust-like spirit of modern science came a debate about the appropriateness of humanity's "playing God."

There was no question, to some, that the supreme example of assuming the role of deity was the creation of something living in a laboratory. In his book *The Nature and Origin of Life: In the Light of New Knowledge* (1906) the French biologist and philosopher Félix Le Dantec (1869–1917) discusses the evolution—or "transformism," the term used in pre-Darwinian discussions in France of species change—of modern species from an early, much simpler organism, "a living protoplasm reduced to the minimum sum of hereditary characters." He wrote, "Archimedes said, in a symbolic proposition which taken literally is absurd: 'Give me a support for a lever and I will move the world.' Just so the Transformist of today has the right to say: Give me a living protoplasm and I will re-make the whole animal and vegetable kingdoms." Le Dantec realized only too well that this task would be hard to achieve with the primitive means at his disposal: "Our acquaintance with colloids [macromolecules] is still so recent and rudimentary that we ought not to count on any speedy success in the efforts to fabricate a living cell." Le Dantec was so certain that the future would bring synthetic cells that he argued, "With the new knowledge acquired by science, the enlightened mind no longer needs to see the fabrication of protoplasm in order to be convinced of the absence of all essential difference and all absolute discontinuity between living and non-living matter."[7]

In the previous century, the boundary between the animate and inanimate had been probed by chemists, including Jöns Jacob Berzelius (1779–1848), a Swedish scientist who is considered one of the pioneers of modern chemistry. Berzelius had pioneered the application of atomic theory to "living" organic chemistry,[8] building on the work of the French father of chemistry, Antoine Lavoisier (1743–1794), and others. He defined the two major branches of chemistry as "organic" and "inorganic"; organic compounds being those that are distinct from all other chemistry by containing carbon atoms. The first-century application of the term "organic" meant "coming from life." But around the time Berzelius came up with the definitions that we still use today in his influential chemistry textbook in the early nineteenth century, the vitalists and neo-vitalists saw the organic world even more uniquely: "Organic substances have at least three constituents . . . they cannot be prepared artificially . . . but only through the affinities associated with vital force. It is made clear that the same rules cannot apply to both organic and inorganic chemistry, the influence of the vital force being essential."[9]

The German chemist Friedrich Wöhler (1800–1882), who worked briefly with Berzelius, has long been credited with a discovery that "disproved" vitalism: the chemical synthesis of urea. You will still find references to his *experimentum crucis* in modern textbooks, lectures, and articles. The achievement was a signal moment in the annals of science, marking the beginning of the end of an influential idea that dated back to antiquity—namely, that there was a "vital force" that distinguished the animate from the inanimate, a distinctive "spirit" that infused all bodies to give them life. From mere chemicals Wöhler seemed to have created something of life itself—a unique moment full of possibilities. With a single experiment, he had transformed chemistry—which, until then, had been divided up into separate domains of life molecules and non-life chemicals—and moved the needle one more notch away from superstition and toward science. His advance came only a decade after Mary Shelley's gothic tale *Frankenstein* was published, itself having appeared only a few years after Giovanni Aldini (1762–1834) attempted to revive a dead criminal with electric shocks.

Wöhler explained his breakthrough in a letter to Berzelius dated January 12, 1828,[10] describing the moment when, at the Polytechnic School in Berlin, he accidentally created urea, the main nitrogen-carrying compound found in the urine of mammals. Wöhler had been attempting to synthesize oxalic acid, a constituent of rhubarb, from the chemicals cyanogen and aqueous ammonia, and ended up with a white crystalline substance. Using careful experimentation, he provided an accurate analysis of natural urea and demonstrated that it had exactly the same composition as his crystals. Until then, urea had only been isolated from animal sources.

Anxious that he had not heard back from Berzelius, Wöhler wrote again, in a letter dated February 22, 1828:

I hope that my letter of January 12th has reached you and although I have been living in a daily or even hourly hope of a reply I will not wait any longer but write to you now because I can no longer, as it were, hold back my chemical urine, and I hope to let out that I can make urea without needing a kidney, whether of man or dog; the ammonium salt of cyanic acid is urea. . . . The supposed ammonium cyanate was easily obtained by reacting lead cyanate with ammonium solution. Silver cyanate and ammonium chloride solution are just as good. Four-sided right-angled prisms, beautifully crystalline, were obtained; when these were treated with acids no cyanic acids were liberated and with alkali no trace of ammonia. But with nitric acid lustrous flakes of an easily crystallized compound, strongly acid in character, were formed; I was disposed to accept this as a new acid for when it was heated neither nitric nor nitrous acid was evolved but a great deal of ammonia. Then I found that if it were saturated with alkali the so-called ammonium cyanate re-appeared and this could be extracted with alcohol. Now, quite suddenly, I had it! All that was needed was to compare urea from urine with this urea from a cyanate.[11]

When Berzelius finally responded, his reaction was both playful and enthusiastic: "After one has begun his immortality in urine, no doubt every reason is present to complete his ascension in the same thing—and truly, Herr Doctor has actually devised a trick that leads down the true path to an immortal name. . . . This will certainly be very enlightening for future theories."

That indeed seemed to be the case. In September 1837 the learned society known as the British Association for the Advancement of Science was addressed in Liverpool by Justus von Liebig (1803–1873), an influential figure who had made key advances in chemistry, such as revealing the importance of nitrogen as a plant nutrient.[12] Von Liebig discussed Wöhler's "extraordinary and to some extent inexplicable production of urea without the assistance of the vital functions," adding that "a new era in science has commenced."[13]

Wöhler's feat soon began to be reported in textbooks, notably in Hermann Franz Moritz Kopp's *History of Chemistry* (1843), which described how it "destroyed the formerly accepted distinction between organic and inorganic bodies." By 1854 the significance of Wöhler's synthesis of urea was underscored when another German chemist, Hermann Kolbe, wrote that it had always been believed that the compounds in animal and plant bodies "owe their formation to a quite mysterious inherent force exclusive to living nature, the so called life force."[14] But now, as a result of Wöhler's "epochal and momentous" discovery, the divide between organic and inorganic compounds had crumbled.

As with the reexamination of many historic events, however, the "revised story" of Wöhler's work can provide new insights that may surprise anyone who accepts the traditional textbook accounts—what the historian of science Peter Ramberg calls the Wöhler Myth. That myth reached its apotheosis in 1937, in Bernard Jaffe's *Crucibles: The Lives and Achievements of the Great Chemists,* a popular history of chemistry that depicted Wöhler as a young scientist who toiled in the "sacred temple" of his laboratory to discredit the mysterious vital force.

Ramberg points out that, given the status of Wöhler's achievement as an experimental milestone, there are surprisingly few known contemporary accounts of the reaction to it. While Berzelius was clearly excited by Wöhler's work, it was not so much in the context of vitalism as it was because the synthesis of urea marked the transformation of a salt-like compound into one that had none of the properties of salt. By showing that ammonium cyanate can become urea through an internal arrangement of its atoms, without gaining or losing in weight, Wöhler had furnished one of the first and best examples of what chemists call isomerism. In doing so he helped to demolish the old view that two bodies that had different physical and chemical properties could not have the same composition.[15]

Historians now generally agree that a single experiment was not responsible for founding the field of organic chemistry. Wöhler's synthesis of urea appears to have had little actual impact on vitalism. Berzelius himself thought that urea, a waste product, was not so much an organic chemical as a substance that occupied the "milieu" between organic and inorganic.[16] Moreover, Wöhler's starting materials had themselves been derived from organic materials, rather than from inorganic ingredients. Nor was his feat unique: four years earlier, he himself had artificially produced another organic compound, oxalic acid, from water and cyanogen.[17] The historian of science John Brooke called the Wöhler synthesis of urea ultimately "no more than a minute pebble obstructing a veritable stream of vitalist thought."

Vitalism, like religion, has not simply disappeared in response to new scientific discoveries. It takes the accumulated weight of evidence from many experiments to displace a belief system. The continual advance of science has progressively staunched vitalism, though the effort has taken centuries, and even today the program to extinguish this mystical belief conclusively is not yet complete.

Some of the key discoveries that *should* have undermined the ancient idea of vitalism date back to 1665, when Robert Hooke (1635–1703), with his pioneering use of a microscope, discovered the first cells. Since his efforts and those of other innovators such as the Dutchman Antonie van Leeuwenhoek (1632–1723), we have accumulated evidence that cells evolved as the primary biological structure for all that we know as life. Vitalism faced more serious challenges with the emergence of modern science during the sixteenth and seventeenth centuries. By 1839, a little over a decade after Wöhler's urea synthesis, Matthias Jakob Schleiden (1804–1881) and Theodor Schwann (1810–1882) wrote, "All living things are composed of living cells." In 1855 Rudolf Virchow (1821–1902), the father of modern pathology, proposed what was called the Biogenic Law: *Omnis cellula e cellula,* or "All living cells arise from pre-existing cells." This stood in marked contrast to the notion of "spontaneous generation," which dates back to the Romans and, as the name suggests, posits that life can arise spontaneously from non-living matter, such as maggots from rotting meat or fruit flies from bananas.

In his famous 1859 experiments Louis Pasteur (1822–1895) disproved spontaneous generation by means of a simple experiment. He boiled broth in two different flasks, one with no cover and open to the air, one with an S-curved top containing a cotton plug. After the flask open to the air cooled, bacteria grew in it, but none grew in the second flask. Pasteur is credited with having proved that microorganisms are everywhere, including the air. As was the case with Wöhler, the full details of his experimental evidence were not as conclusive as has often been portrayed, and it would take the subsequent work of German scientists to provide definitive proof.[18]

Pasteur's experiments led some subsequent scientists to rule out the possibility that life had originally developed from, or could be developed out of, inorganic chemicals. In 1906 the French biologist and philosopher Félix Le Dantec wrote, "It is often said that Pasteur demonstrated the uselessness of such efforts as . . . men of science endeavoring to reproduce life in their laboratories. Pasteur showed only this: By taking certain precautions we can keep all invasion on the part of living species actually existing in certain substances which might serve them as food. And that is all. The problem of protoplasm synthesis remains what it was."[19]

Although Pasteur had shown how to exclude life from a sterile environment, he had not advanced our understanding of how, over billions of years, life had become established on the infant Earth. In 1880 the German evolutionary biologist August Weismann (1834–1914) introduced an important corollary to the Biogenic Law which pointed back to the ultimate origin: "Cells living today can trace their ancestry back to ancient times." In other words, there must be a common ancestral cell. And that, of course, takes us to Charles Darwin's revolutionary work, *On the Origin of Species* (1859). Darwin (1809–1882), along with the British naturalist and explorer Alfred Russel Wallace (1823–1913), argued that there exists within all creatures' variations or changes in the species characteristics that are passed down through the generations. Some variations result in advantageous forms that thrive with each successive generation, so they—and their genes—become more common. This is natural selection. In time, as novel versions accumulate, a lineage may evolve to such an extent that it can no longer exchange genes with others that were once its kin. In this way, a new species is born.

Despite such scientific advances, vitalism had passionate advocates into the twentieth century. Among them was Hans Driesch (1867–1941), an eminent German embryologist who, because the intellectual problem of the formation of a body from a patternless single cell seemed to him otherwise insoluble, had turned to the idea of entelechy (from the Greek *entelécheia*), which requires a "soul," "organizing field," or "vital function" to animate the material ingredients of life. In 1952 the great British mathematician Alan Turing would show how a pattern could emerge in an embryo *de novo*.[20] Likewise, the French philosopher Henri-Louis Bergson (1874–1948) posited an *élan vital* to overcome the resistance of inert matter in the formation of living bodies. Even today, although most serious scientists believe vitalism to be a concept long since disproven, some have not abandoned the notion that life is based on some mysterious force. Perhaps this should not come as a surprise: the word *vitalism* has always had as many meanings as it has had supporters, and a widely accepted definition of life remains elusive.

In our own time a new kind of vitalism has emerged. In this more refined form the emphasis is not so much on the presence of a vital spark as on how current reductionist, materialist explanations seem inadequate to explain the mystery of life. This line of thought reflects the belief that the complexity of a living cell arises out of vast numbers of interacting chemical processes forming interconnected feedback cycles that cannot be described merely in terms of those component processes and their constituent reactions. As a result, vitalism today manifests itself in the guise of shifting emphasis away from DNA to an "emergent" property of the cell that is somehow greater than the sum of its molecular parts and how they work in a particular environment.

This subtle new vitalism results in a tendency by some to downgrade or even ignore the central importance of DNA. Ironically, reductionism has not helped. The complexity of cells, together with the continued subdivision of biology into teaching departments in most universities, has led many down the path of a protein-centric versus a DNA-centric view of biology. In recent years, the DNA-centric view has seen an increasing emphasis on epigenetics, the system of "switches" that turns genes on and off in a cell in response to environmental factors such as stress and nutrition. Many now behave as if the field of epigenetics is truly separate from and independent of DNA-driven biology. When one attributes unmeasurable properties to the cell cytoplasm, one has unwittingly fallen into the trap of vitalism. The same goes for the emphasis of the mysterious emergent properties of the cell over DNA, which is tantamount to a revival of *Omnis cellula e cellula,* the idea that all living cells arise from pre-existing cells.

It is certainly true that cells have evolved as the primary biological foundation for all that we know as life. Understanding their structure and content has, as a result, been the basis for the important central disciplines of cell biology and biochemistry/metabolism. However, as I hope to make clear, cells will die in minutes to days if they lack their genetic information system. The longest exception to this are our red blood cells that have a half-life of 120 days. Without genetic information cells have no means to make their protein components or their envelope of lipid molecules, which form the membrane that holds their watery contents. They will not evolve, they will not replicate, and they will not live.

Despite our recognition that the myth that has obscured Wöhler's synthesis of urea does not accurately reflect the historical facts of the case, the fundamental logic of his experiment still exerts a powerful and legitimate influence on scientific methods. Today it is standard practice to prove a chemical structure is correct by undertaking that chemical's synthesis and demonstrating that the synthetic version has all the properties of a natural product. Tens of thousands of scientific papers start with this premise or contain the phrase "proof by synthesis." My own research has been guided by the principles of Wöhler's 1828 letter. When in May 2010 my team at the J. Craig Venter Institute (JCVI) synthesized an entire bacterial chromosome from computer code and four bottles of chemicals, then booted up the chromosome in a cell to create the first synthetic organism, we drew parallels to the work of Wöhler[21] and his "synthesis as proof."

The materialistic view of life as machines has led some to attempt the creation of artificial life outside of biology, with mechanical systems and mathematical models. By the 1950s, when DNA was finally becoming accepted as the genetic material, the mechanistic approach had already been aired in the scientific literature. In this version, life would arise from complex *mechanisms*, rather than complex *chemistry*. In 1929 the young Irish crystallographer John Desmond Bernal (1901–1971) imagined the possibility of machines with a lifelike ability to reproduce themselves, in a "post-biological future" he described in *The World, the Flesh & the Devil*: "To make life itself will be only a preliminary stage. The mere making of life would only be important if we intended to allow it to evolve of itself anew."

A logical recipe to create these complex mechanisms was developed in the next decade. In 1936 Alan Turing, the cryptographer and pioneer of artificial intelligence, described what has come to be known as a Turing machine, which is described by a set of instructions written on a tape. Turing also defined a universal Turing machine, which can carry out any computation for which an instruction set can be written. This is the theoretical foundation of the digital computer.

Turing's ideas were developed further in the 1940s, by the remarkable American mathematician and polymath John von Neumann, who conceived of a self-replicating machine. Just as Turing had envisaged a universal machine, so von Neumann envisaged a universal constructor. The Hungarian-born genius outlined his ideas in a lecture, "The General and Logical Theory of Automata," at the 1948 Hixon Symposium, in Pasadena, California. He pointed out that natural organisms "are, as a rule, much more complicated and subtle, and therefore much less well understood in detail than are artificial automata"; nevertheless, he maintained that some of the regularities we observe in the former might be instructive in our thinking about and planning of the latter.

Von Neumann's machine includes a "tape" of cells that encodes the sequence of actions to be performed by it. Using a writing head (termed a "construction arm") the machine can print out (construct) a new pattern of cells, enabling it to make a complete copy of itself, and the tape. Von Neumann's replicator was a clunky-looking structure consisting of a basic box of eighty by four hundred squares, the constructing arm, and a "Turing tail," a strip of coded instructions consisting of another one hundred and fifty thousand squares. ("[Turing's] automata are purely computing machines," explained von Neumann. "What is needed . . . is an automaton whose output is other automata."[22]) In all, the creature consisted of about two hundred thousand such "cells." To reproduce, the machine used "neurons" to provide the logical control, transmission cells to carry messages from the control centers, and "muscles" to change the surrounding cells. Under the instructions of the Turing tail, the machine would extend the arm, and then scan it back and forth, creating a copy of itself by a series of logical manipulations. The copy could then make a copy, and so on and so forth.

The nature of those instructions became clearer as the digital world and the biological worlds of science advanced in parallel during this period. Erwin Schrödinger wrote then what appears to be the first reference to his "code-script": "It is these chromosomes, or probably only an axial skeleton fiber of what we actually see under the microscope as the chromosome, that contain in some kind of code-script the entire pattern of the individual's future development and of its functioning in the mature state." Schrödinger went on to state that the "code-script" could be as simple as a binary code: "Indeed, the number of atoms in such a structure need not be very large to produce an almost unlimited number of possible arrangements. For illustration, think of the Morse code. The two different signs of dot and dash in well-ordered groups of not more than four allow of thirty different specifications."[23]

Even though von Neumann conceived his self-replicating automaton some years before the actual hereditary code in the DNA double helix was discovered, he did lay stress on its ability to evolve. He told the audience at his Hixon lecture that each instruction that the machine carried out was "roughly effecting the functions of a gene" and went on to describe how errors in the automaton "can exhibit certain typical traits which appear in connection with mutation, lethally as a rule, but with a possibility of continuing reproduction with a modification of traits." As the geneticist Sydney Brenner has remarked, it can be argued that biology offers the best real-world examples of the machines of Turing and von Neumann: "The concept of the gene as a symbolic representation of the organism—a code script—is a fundamental feature of the living world."[24]

Von Neumann followed up on his original notion of a replicator by conceiving of a purely logic-based automaton, one that did not require a physical body and a sea of parts but was based instead on the changing states of the cells in a grid. His colleague at Los Alamos, New Mexico (where they worked on the Manhattan project), Stanislaw Ulam, had suggested that von Neumann develop his design using a mathematical abstraction, such as the one Ulam himself had used to study crystal growth. Von Neumann unveiled the resulting "self-reproducing automaton"—the first cellular automaton—at the Vanuxem Lectures on "Machines and Organisms" at Princeton University, New Jersey, between March 2 and 5, 1953.

While efforts continued to model life, our understanding of the actual biology underlying it changed when, on April 25, 1953, James Watson and Francis Crick published a milestone paper in the journal *Nature*,[25] "Molecular Structure of Nucleic Acids: A Structure for Deoxyribose Nucleic Acid." Their study, based in Cambridge, England, proposed the double helical structure of DNA, based on X-ray crystal data obtained by Rosalind Franklin and Raymond Gosling at King's College London. Watson and Crick described the elegantly functional molecular structure of the double helix, and how DNA is reproduced so its instructions can be passed down the generations. This is nature's self-reproducing automaton.

The onset of efforts to create another kind of self-reproducing automaton, along with the beginnings of artificial-life research, date back to around this period, when the first modern computers came into use. The discovery of the coded nature of life's genetic information system led naturally to parallels with Turing machines. Turing himself, in his key 1950 paper on artificial intelligence, discussed how survival of the fittest was "a slow method" that could possibly be given a boost, not least because an experimenter was not restricted to random mutations.[26] Many began to believe that artificial life would emerge from complex logical interactions within a computer.

Various streams of thought combined at this point: the theories of von Neumann, with his work on early computers and his self-reproducing automaton; of Turing, who posed basic questions about machine intelligence[27]; and of the American mathematician Norbert Weiner, who applied ideas from information theory and self-regulating processes to living things in the field of cybernetics,[28] described in his book *Cybernetics*, published in 1948. There were subsequently many notable attempts to kindle life in a computer. One of the earliest took place at the Institute for Advanced Study in Princeton, New Jersey, in 1953, when the Norwegian-Italian Nils Aall Barricelli, a viral geneticist, carried out experiments "with the aim of verifying the possibility of an evolution similar to that of living organisms taking place in an artificially created universe."[29] He reported various "biophenomena," such as the successful crossing between parent "organisms," the role of sex in evolutionary change, and the role of cooperation in evolution.[30]

Perhaps the most compelling artificial life experiment took place several decades later, in 1990, when Thomas S. Ray, at the University of Delaware, programmed the first impressive attempt at Darwinian evolution inside a computer, in which organisms—segments of computer code—fought for memory (space) and processor power (energy) within a cordoned-off "nature reserve" inside the machine. To achieve this he had to overcome a key obstacle: programming languages are "brittle," in that a single mutation—a line, letter, or point in the wrong place—brings them to a halt. Ray introduced some changes that made it much less likely that mutations could disable his program. Other versions of computer evolution followed, notably Avida,[31] software devised by a team at Caltech in the early 1990s to study the evolutionary biology of self-replicating computer programs. Researchers believed that with greater computer power, they would be able to forge more complex creatures—the richer the computer's environment, the richer the artificial life that could go forth and multiply.

Even today, there are those, such as George Dyson, in his book *Turing's Cathedral* (2012), who argue that the primitive slivers of replicating code in Barricelli's universe are the ancestors of the multi-megabyte strings of code that replicate in the digital universe of today, in the World Wide Web and beyond.[32] He points out that there is now a cosmos of self-reproducing digital code that is growing by trillions of bits per second, "a universe of numbers with a life of their own."[33] These virtual landscapes are expanding at an exponential rate and, as Dyson himself has observed, are starting to become the digital universe of DNA.

But these virtual pastures are, in fact, relatively barren. In 1953, only six months after he had attempted to create evolution in an artificial universe, Barricelli had found that there were significant barriers to be overcome in any attempt to generate artificial life in the computer. He reported that "something is missing if one wants to explain the formation of organs and faculties as complex as those of living organisms . . . No matter how many mutations we make, the numbers will always remain numbers. Numbers alone will never become living organisms!"[34]

Artificial life as originally conceived has had a new virtual life in the form of games and movies, with the murderous Hal 9000 of *2001: A Space Odyssey*, the genocidal Skynet of the *Terminator* films, and the malevolent machines of *The Matrix*. However, the reality still lags far behind. In computer-based artificial life there is no distinction between the genetic sequence or genotype of the manufactured organism and its phenotype, the physical expression of that sequence. In the case of a living cell, the DNA code is expressed in the form of RNA, proteins and cells, which form all of the physical substances of life. Artificial life systems quickly run out of steam, because genetic possibilities within a computer model are not open-ended but predefined. Unlike in the biological world, the outcome of computer evolution is built into its programming.

In science, the fields of chemistry, biology, and computing have come together successfully in my own discipline of genomics. Digital computers designed by DNA machines (humans) are now used to read the coded instructions in DNA, to analyze them and to write them in such a way as to create new kinds of DNA machines (synthetic life). When we announced our creation of the first synthetic cell, some had asked whether we were "playing God." In the restricted sense that we had shown with this experiment how God was unnecessary for the creation of new life, I suppose that we were. I believed that with the creation of synthetic life from chemicals, we had finally put to rest any remaining notions of vitalism once and for all. But it seems that I had underestimated the extent to which a belief in vitalism still pervades modern scientific thinking. Belief is the enemy of scientific advancement. The belief that proteins were the genetic material set back the discovery of DNA as the information-carrier, perhaps by as much as half a century.

During the latter half of the twentieth century we came to understand that DNA was Schrödinger's "code-script," deciphered its complex message, and began to figure out precisely how it guides the processes of life. This epic adventure in understanding would mark the birth of a new era of science, one that lay at the nexus of biology and technology.

3 Dawn of the Digital Age of Biology

If we are right, and of course that is not yet proven, then it means that nucleic acids are not merely structurally important but functionally active substances in determining the biochemical activities and specific characteristics of cells and that by means of a known chemical substance it is possible to induce predictable and hereditary changes in cells. This is something that has long been the dream of geneticists.

—Oswald Avery, in a letter to his brother Roy, 1943[1]

It was in the same year that Schrödinger delivered his milestone lectures in Dublin that the chemical nature of his "code-script" and of all inheritance was revealed at last, providing new insights into a subject that has obsessed, fascinated, bemused, and confused our ancestors from the very dawn of human consciousness. A great warrior has many children, yet none of them has either the build or the inclination for battle. Some families are affected by a particular type of illness, yet it ripples down the generations in an apparently haphazard way, affecting one descendant but not another. Why do certain physical features of parents and even more distant relatives appear or, perhaps more puzzlingly, not appear in individuals? For millennia, the same questions have been asked, not only of our own species but of cattle, crops, plants, dogs, and so on.

Many insights about these mysteries have emerged since the birth of agriculture and the domestication of animals millennia ago. Aristotle had a vague grasp of the fundamental principles when he wrote that the "concept" of a chicken is implicit in a hen's egg, that an acorn is "informed" by the arrangement of an oak tree. In the eighteenth century, as a result of the rise of knowledge of plant and animal diversity along with taxonomy, new ideas about heredity began to appear. Charles Darwin's grandfather, Erasmus Darwin (1731–1802), a formidable intellectual force in eighteenth-century England, formulated one of the first formal theories of evolution in the first volume of *Zoonomia; or the Laws of Organic Life*[2] (1794–1796), in which he stated that "all living animals have arisen from one living filament." Classical genetics, as we understand it, has its origin in the 1850s and 1860s, when the Silesian friar Gregor Mendel (1822–1884) attempted to draw up the rules of inheritance governing plant hybridization. But it is only in the past seventy years that scientists have made the remarkable discovery that the "filament" that Erasmus Darwin proposed is in fact used to program every organism on the planet with the help of molecular robots.

Until the middle of the last century most scientists believed that only proteins carried genetic information. Given that life is so complex, it was thought that DNA, a polymer consisting of only four chemical units, was far too simple in composition to transmit enough data to the next generation, and was merely a support structure for genetic protein material. Proteins are made of twenty different amino acids and have complex primary, secondary, tertiary, and quaternary structures, while DNA is a polymer thread. Only proteins seemed sufficiently complex to function as Schrödinger's "aperiodic crystal," able to carry the full extent of information that must be transferred from cell to cell during cell division.

That attitude would begin to change in 1944, when details of a beautiful, simple experiment were published. The discovery that DNA, not protein, was the actual carrier of genetic information was made by Oswald Avery (1877–1955), at Rockefeller University, New York. By isolating a substance that could transfer some of the properties of one bacterial strain to another through a process called transformation, he discovered that the DNA polymer was actually what he called the "transforming factor" that endowed cells with new properties.

Avery, who was then sixty-five and about to retire, along with his colleagues Colin Munro MacLeod and Maclyn McCarty, had followed up a puzzling observation made almost two decades earlier by the bacteriologist Frederick Griffith (1879–1941), in London. Griffith had been studying the bacterium pneumococcus (*Streptococcus pneumoniae*), which causes pneumonia epidemics and occurs in two different forms: an R form, which looks rough under the microscope and is not infectious, and an S, or smooth, form, which is able to cause disease and death. Both R and S forms are found in patients with pneumonia.

Griffith wondered if the lethal and benign forms of the bacteria were interconvertible. To answer this question, he devised a clever experiment in which he injected mice with the noninfective R cells along with S cells that he had killed with heat. One would have expected that the mice would survive, since when the virulent S form was killed and it alone introduced, the rodents lived. Unexpectedly, however, the mice died when the living, avirulent R form accompanied the dead S cells. Griffith recovered both live R and S cells from the dead mice. He reasoned that some substance from the heat-killed S cells was transferred to the R cells to turn them into the S type. Since this change was inherited by subsequent generations of bacteria, it was assumed that the factor was genetic material. He called the process "transformation," though he had no idea about the true nature of the "transforming factor."

The answer would come almost twenty years later when Avery and his colleagues repeated Griffith's experiment and proved by a process of elimination that the factor was DNA. They had progressively removed the protein, RNA, and DNA using enzymes that digest only each individual component of the cell: in this case, proteases, RNases, and DNases, respectively.[3] The impact of their subsequent paper was far from instantaneous, however, because the scientific community was slow to abandon the belief that the complexity of proteins was necessary to explain genetics. In *Nobel Prizes and Life Sciences* (2010), Erling Norrby, former secretary general of the Royal Swedish Academy of Sciences, discusses the reluctance to accept Avery's discovery, for while his team's work was compelling, skeptics reasoned that there was still a possibility that minute amounts of some other substance, perhaps a protein that resisted proteases, was responsible for the trans-formation.[4]

Great strides continued to be made in understanding proteins, no-tably in 1949, when Briton Frederick Sanger determined the sequence of amino acids in the hormone insulin, a remarkable feat that would be rewarded with a Nobel Prize. His work showed that proteins were not combinations of closely related substances with no unique structure but were indeed a single chemical.[5] Sanger, for whom I have the great-est respect, is without doubt one of the most masterful science innova-tors of all time, due to his emphasis on developing new techniques.[6] ("Of the three main activities involved in scientific research, thinking, talking, and doing, I much prefer the last and am probably best at it. I am all right at the thinking, but not much good at the talking."[7]) His approach paid handsome dividends.

The idea that nucleic acids hold the key to inheritance gradually be-gan to take hold in the late 1940s and early 1950s, as other successful transformation experiments were performed—for example, the RNA from tobacco mosaic virus was shown to be infectious on its own. Still, recognition that DNA was the genetic material was slow to come. The true significance of the experiments by Avery, MacLeod, and McCarty only became clear as data accumulated over the next decade. One key piece of support came in 1952, when Alfred Hershey and Martha Cowles Chase demonstrated that DNA was the genetic material of a virus known as the T2 bacteriophage, which is able to infect bacteria.[8] The understanding that DNA was the genetic material received a big boost in 1953, when its structure was revealed by Watson and Crick, while working in Cambridge, England. Earlier studies had established that DNA is composed of building blocks called nucleotides, consist-ing of a deoxyribose sugar, a phosphate group, and four nitrogen bases—adenine (A), thymine (T), guanine (G), and cytosine (C). Phos-phates and sugars of adjacent nucleotides link to form a long polymer. Watson and Crick determined how these pieces fit together in an ele-gant three-dimensional structure.

To achieve their breakthrough they had used critical data from other scientists. From Erwin Chargaff, a biochemist, they learned that the four different chemical bases in DNA are to be found in pairs, a critical insight when it came to understanding the "rungs" down the ladder of life. (A part of the History of Science collection at my not-for-profit J. Craig Venter Institute is Crick's lab notebook from this time, recording his unsuccessful attempts to repeat Chargaff's experiment.) From Maurice Wilkins, who had first excited Watson with his pioneering X-ray studies of DNA, and Rosalind Franklin they obtained the key to the solution. It was Wilkins who had shown Watson the best of Franklin's X-ray photographs of DNA. The photo numbered fifty-one (also part of the collection at the Venter Institute), taken by Raymond Gosling in May 1952, revealed a black cross of reflections and would prove the key to unlocking the molecular structure of DNA, revealing it to be a double helix, where the letters of the DNA code corresponded to the rungs.[9]

On April 25, 1953, Watson and Crick's article "Molecular Structure of Nucleic Acids: A Structure for Deoxyribose Nucleic Acid"[10] was published in *Nature*. The helical DNA structure came as an epiphany, "far more beautiful than we ever anticipated," explained Watson, because the complementary nature of the letters—component nucleotides—of DNA (the letter A always pairs with T, and C with G) instantly revealed how genes were copied when cells divide. While this was the long-sought mechanism of inheritance, the response to Watson and Crick's paper was far from instantaneous. Recognition eventually did come, and nine years later Watson, Crick, and Wilkins would share the 1962 Nobel Prize in Physiology or Medicine "for their discoveries concerning the molecular structure of nucleic acids and its significance for information transfer in living material."

The two scientists who supplied the key data were, however, not included: Erwin Chargaff was left embittered,[11] and Rosalind Franklin had died in 1958, at the age of 37, from ovarian cancer. Although Oswald Avery had been nominated several times for the Nobel Prize, he died in 1955 before acceptance of his accomplishments was sufficient for it to be awarded him. Erling Norrby quotes Göran Liljestrand, secretary of the Nobel Committee of the Karolinska Institute, from his 1970 summary of Nobel Prizes in Physiology or Medicine: "Avery's discovery in 1944 of DNA as the carrier of heredity represents one of the most important achievements in genetics, and it is to be regretted that he did not receive the Nobel Prize. By the time dissident voices were silenced, he had passed away."[12]

Avery's story illustrates that, even in the laboratory, where the rational, evidence-based view of science should prevail, belief in a particular theory or hypothesis can blind scientists for years or even decades. Avery's, MacLeod's, and McCarty's experiments were so simple and so elegant that they could have easily been replicated; it remains puzzling to me that this had not been done earlier. What distinguishes science from other fields of endeavor is that old ideas fall away when enough data accumulates to contradict them. But, unfortunately, the process takes time.

Cellular life is in fact dependent on two types of nucleic acid: deoxyribonucleic acid, DNA, and ribonucleic acid, RNA. Current theory is that life began in an RNA world, because it is more versatile than DNA. RNA has dual roles as both an information carrier and as an enzyme (ribozyme), being able to catalyze chemical reactions. Like DNA, RNA consists of a linear string of chemical letters. The letters are represented by A, C, G, and either T in DNA or U in RNA. C always binds to G; A binds only to T or U. Just like DNA, a single strand of RNA can bind to another strand consisting of complementary letters. Watson and Crick proposed that RNA is a copy of the DNA message in the chromosomes and takes the message to the ribosomes, where proteins are manufactured. The DNA software is "transcribed," or copied, into the form of a messenger RNA (mRNA) molecule. In the cytoplasm, the mRNA code is "translated" into proteins.

It wasn't until the 1960s that DNA was finally widely recognized as "the" genetic material, but it would take the work of Marshall Warren Nirenberg (1927–2010), at the National Institutes of Health, Bethesda, Maryland, and India-born Har Gobind Khorana (1922–2011), of the

University of Wisconsin, Madison, to actually decipher the genetic code by using synthetic nucleic acids. They found that DNA uses its four different bases in sets of three—called codons—to code for each of the twenty different amino acids that are used by cells to make proteins. This triplet code therefore has sixty-four possible codons, some of which serve as punctuation (stop codons) to signal the end of a protein sequence. Robert W. Holley (1922–1993), of Cornell, elucidated the structure of another species of RNA, called transfer RNA (tRNA), which carries the specified amino acids to the spectacular molecular machine called a ribosome, where they are assembled into proteins. For these illuminating studies, Nirenberg, Khorana, and Holley shared the Nobel Prize in 1968.

I had the privilege of meeting all three men at various times but got to know Marshall Nirenberg particularly well while I was working at the National Institutes of Health. Nirenberg's lab and office were one floor below mine, in Building 36 on the sprawling NIH campus, and I visited him often during my early days of DNA sequencing and genomics. A genial man, deeply interested in all areas of science, he was always excited about new technology, right up until the time of his death. His discovery of the genetic code with Khorana will be remembered as one of the most significant in all of bioscience, as it explained how the linear DNA polymer codes for the linear polypeptide sequence of proteins. This is the core principle of the "central dogma" of molecular biology: information travels from the nucleic acid to the proteins.

The 1960s were the start of the molecular-biology revolution due in part to the ability to splice DNA using restriction enzymes. Restriction enzymes were independently discovered by Werner Arber, in Geneva, and Hamilton O. Smith, working in Baltimore. "Ham" Smith, a longtime friend and collaborator, published two important papers in 1970 describing a restriction enzyme isolated from the bacterium *Haemophilus influenzae*. One of the key biochemical mechanisms used by bacteria to protect themselves from foreign DNA are enzymes that can rapidly chop up DNA from other species that have entered the cell, always cutting its strands at one particular defined sequence of code and no other. Daniel Nathans worked with Smith in Baltimore to pioneer the application of restriction enzymes to genetic fingerprinting and mapping. The enzymes enable scientists to manipulate DNA just as one uses a word processor to cut and paste text. The ability to cut genetic material precisely at known sites is the basis of all genetic engineering and DNA fingerprinting. The latter has revolutionized forensic science and criminal identification from DNA left at crime scenes, in the form of fingerprints, hair, skin, semen, and saliva, for example. Smith, Nathans, and Arber would share a Nobel Prize in 1978 for their discoveries; without them, the field of molecular engineering might not exist.

The 1970s brought the beginning of the gene-splicing revolution, a development potentially as revolutionary as the birth of agriculture in the Neolithic Era. When DNA from one organism is artificially introduced into the genome of another and then replicated and used by that other organism, it is known as recombinant DNA. The invention of this technology was largely the work of Paul Berg, Herbert Boyer, and Stanley Norman Cohen. Working at Stanford, Berg began to wonder whether it would be possible to insert foreign genes into a virus, thereby creating a "vector" that could be used to carry genes into new cells. His landmark 1971 experiment involved splicing a segment of the DNA of a bacterial virus, known as lambda, into the DNA of a monkey virus, SV40.[13]

Berg would share the 1980 Nobel Prize in Chemistry for his work, but he did not take the next step of introducing recombinant DNA into animals. The first transgenic mammal was created in 1974 by Rudolf Jaenisch and Beatrice Mintz, who inserted foreign DNA into mouse embryos.[14] Because of the growing public unease over the potential dangers of such experimentation, Berg played an active role in debating to what degree such studies should be constrained and limited. In 1974 a group of American scientists recommended a moratorium on this research. Voluntary guidelines were drawn up at a highly influential meeting organized the following year by Berg at the Asilomar Conference Grounds, in Pacific Grove, California. The fear of some was that recombinant organisms might have unexpected consequences, such as causing illness or death, and that they might escape the laboratory and spread. This concern was balanced by arguments in support of the potential of genetic engineering, notably those of Joshua Lederberg, a Stanford professor and Nobel laureate.[15] In 1976 the National Institutes of Health issued its own guidelines for the safe conduct of recombinant-DNA research, the repercussions of which are still being felt in the ongoing debates about genetically altered crops and the more recent discussion about the use and misuse of research on the genetics of influenza.

After Berg's 1971 gene-splicing experiment, the next advance in molecular cloning was the insertion of DNA from one species of bacterium into another, where it would replicate every time the bacterium divided. This step was taken in 1972 by Boyer, at the University of California at San Francisco, working with Cohen, of Stanford University. Their research, in which the DNA from *Staphylococcus* was propagated in *E. coli*, established that genetic materials could indeed be transferred between species, thereby disproving a long-held belief. An even greater triumph of interspecies cloning was marked by the insertion into *E. coli* of genes from the South African clawed frog *Xenopus*, a favorite experimental animal. Despite public unease, a number of companies were rapidly created to exploit recombinant-DNA technology.

At the forefront of the biotechnology revolution was the company Genentech, founded in 1976 by Boyer and venture capitalist Robert A. Swanson. The following year, before Genentech had even moved into its own facilities, Boyer and Keiichi Itakura, at the City of Hope medical center, in Duarte, California, working with Arthur Riggs, had used recombinant-DNA technology to produce a human protein called somatostatin (which plays a major role in regulating the growth hormone) in *E. coli*. After this milestone they turned to the more complicated insulin molecule, for which a huge potential market existed in replacing the pig insulin then being used for the treatment of diabetes. Eli Lilly and Company signed a joint-venture agreement with

Genentech to develop the production process, and in 1982 the recombinant insulin protein, under the brand name Humulin, became the first biotechnology product to appear on the market. By then Genentech had many rivals, including a number of small start-ups backed by major pharmaceutical companies.

Molecular biology has grown explosively from these early discoveries to a field that is now practiced at every university worldwide and is the basis for a multibillion-dollar business manufacturing kits, tests and reagents, and scientific instruments. Genes from almost every species, including bacteria, yeast, plants, and mammals, have been or are being cloned and studied on a daily basis. Metabolic pathways are being engineered in research laboratories and in biotech companies to coax cells into generating products ranging from pharmaceuticals to food and industrial chemicals to energy molecules.

Parallel to this explosion in understanding the DNA software of life has been the substantial progress in describing the protein hardware of life. Proteins are the basic building blocks of the cell, the fundamental structural unit of all known living entities, from a single bacterium to the one hundred trillion cells that make up the human body. As mentioned above, the world of the cell was first revealed by Robert Hooke, whom some refer to as England's Leonardo da Vinci. Hooke was the earliest major British figure to show how the experimental method, using instruments, actually works and produces progressive knowledge. In his masterful *Micrographia*[16] (1665) Hooke described cells (the word "cell" itself comes from the Latin *cellula*, a small room), after he had viewed the honeycombed structure of sliced cork through his microscope. Each and every living entity on Earth has a basic cellular structure enshrouded by a membrane that creates an isolated interior volume. That interior holds the genetic material and the cellular machinery for its replication.

In the first two decades of the twentieth century the effort to identify the molecular basis of that hardware by the field of microbiology was dominated by what was called "colloidal theory." At that time there was no clear-cut evidence for the existence of large molecules, and the "biocolloidists" argued that antibodies, enzymes, and the like in fact consisted of colloids, mixtures of varying compositions of little molecules.[17] They put emphasis not on giant organic molecules held together by strong covalent bonds but on aggregates of small molecules held together by relatively weak bonds. During the early 1920s, however, that view was shaken by the German organic chemist Hermann Staudinger (1881–1965), who showed that large molecules such as starch, cellulose, and proteins are in fact long chains of short, repeating molecular units held together by covalent bonds. However, Staudinger's notion of what he called *Makromoleküle* (macromolecules) was at first almost universally opposed. Macromolecular theory was even rejected by Staudinger's colleagues at the Eidgenössische Technische Hochschule (ETH), in Zürich, where he was a professor until he moved to Freiburg, in 1926. It was only in 1953 (the year of the double-helix discovery) that Staudinger was eventually awarded the Nobel Prize for his important contribution.

In recent years we have come to regard that basic unit of life, the cell, as a factory, an interlocking series of assembly lines run by protein machines[18] that have evolved over thousands, millions, or even billions of years to carry out specific tasks. This model marks a resurgence of an idea that was current in the seventeenth century, notably though the efforts of Marcello Malpighi (1628–1694), an Italian doctor who conducted pioneering studies with microscopes.[19] Malpighi proposed that minute "organic machines" controlled bodily functions.

Today we know that there are many well-characterized classes of protein. Catalysts, for example, speed a dizzying array of chemical reactions, while fibrous proteins, such as collagen, are a major structural element, accounting for one quarter of all the proteins found in vertebrates, backboned creatures including mammals. Elastin, which resembles rubber, is the basis of lung and arterial-wall tissue. The membranes around our cells contain proteins that help move molecules into and out of the interior and are involved in cell communication; globular proteins bind, transform, and release chemicals. And so on.

DNA sequence directly codes each protein's structure, which determines its activity. The genetic code defines the linear sequence of amino acids, which in turn determines the complex three-dimensional structure of the final protein. After synthesis, this linear polypeptide sequence folds into the proper characteristic shape: some parts form sheets, while others stack, loop, curl, and twist into spirals (helices) and other complicated configurations that define the workings of the machine. Some parts of the protein machine bend, while others are rigid. Some proteins are subassemblies, parts of a greater three-dimensional protein machine.

Let's look at ATP synthase as one remarkable, and energetic, example of a molecular machine. This enzyme, some two hundred thousand times smaller than a pinhead, is made of thirty-one proteins and, as it rotates about sixty times per second, is able to create the energy currency of cells, a molecule called adenosine triphosphate, or ATP. You would not be able to move, think, or breathe without this machine. Other proteins are motors, such as dynein, which enables sperm to wriggle; myosin, which moves muscles; and kinesin, which "walks" on a pair of feet (as the ATP fuel docks, one foot swings out and flaps about before latching on to make its next step) and has a tail to carry cargo around in cells. Some of these transport robots are customized to carry only one kind of cargo: among them is hemoglobin, which is composed of four protein chains, two alpha chains and two beta chains, each of which possesses a ring-like heme group that has an iron atom at its heart to cart oxygen around the body. Iron would usually cling tightly to oxygen, but this machine has evolved to ensure that the oxygen molecule binds reversibly at the four-heme sites in each hemoglobin molecule.

Light-absorbing pigment is the secret of one of the most important machines of all, the one that drives the living economy of the oceans and surface of the planet. While different species of plants, algae, and bacteria have evolved different mechanisms to harvest light energy, they all share a molecular feature known as a photosynthetic reaction center. There one finds antenna proteins, which are made up of multiple light-absorbing chlorophyll pigments. They capture sunlight in the form of particles of light called photons, and then transfer their energy through a series of molecules to the reaction center, where it is used to convert carbon dioxide into sugars with great efficiency. Photosynthetic processes take place in spaces so tightly packed with pigment molecules that quantum-mechanical effects come into play.[20] (The most head-spinning branch of physics, quantum mechanics—established by Erwin Schrödinger and many others—deals with phenomena at microscopic scales.) This is one of several quantum machines used by living things in vision, electron- and proton-tunneling, olfactory sensing, and magnetoreception.[21] This extraordinary finding is another testament to the insights of Schrödinger, who had also considered the possibility that quantum fluctuations had a role in biology.[22]

Each molecular machine has evolved to carry out a very specific task, from recording visual images to flexing muscles, and to do it automatically. That is why one can think of them as little robots. As Charles Tanford and Jacqueline Reynolds write in *Nature's Robots* (2001), "it doesn't have consciousness; it doesn't have a control from the mind or higher center. Everything a protein does is built into its linear code, derived from the DNA code."

The most important breakthrough in molecular biology, outside of the genetic code, was in determining the details of the master robot—the ribosome—that carries out protein synthesis and so directs the manufacture of all other cellular robots. Molecular biologists have known for decades that the ribosome is at the focus of the choreography of protein manufacture. To function, the ribosome needs two things: a messenger RNA (mRNA) molecule, which has copied the instructions for making a protein from the storehouse of DNA genetic information in the cell; and transfer RNA (tRNA), which carries on its back the amino acids used to make the protein. The ribosome reads the mRNA sequence, one codon at a time, and matches it to the anti-codon on each tRNA, lining up their cargo of amino acids in the proper order. The ribosome also acts as a catalyst, a ribozyme, and fuses the amino acids with a covalent chemical bond to add to the growing protein chain. Synthesis is terminated when the RNA sequence codes for a "stop," and the polymer of amino acids must then fold into its required three-dimensional structure to be a biologically active protein.

Bacterial cells contain as many as a thousand ribosome complexes, which enable continuous protein synthesis, both to replace degraded proteins and to make new ones for daughter cells during cell division. One can study a ribosome under an electron microscope and watch it bend and deform as it works. At a key point in the protein synthesis process there is a ratcheting rotation deep within it.[23] Overall, protein synthesis is extremely fast, requiring only seconds to make chains of one hundred or so amino acids.

As was the case with the double helix, X-ray crystallography was needed to reveal the ribosome's detailed structure. First, however, someone had to make the ribosomes crystallize—like salt in a solution crystallizes when the water evaporates—to leave well-organized crystals with millions of ribosomes assembled into regular patterns that could be studied with X-rays. A key advance came in the 1980s, when Ada E. Yonath, in Israel, collaborated with Heinz-Günter Wittmann, in Berlin, to grow crystals from bacterial ribosomes, isolated from microorganisms from hot springs and the Dead Sea. The secrets of the bacterial ribosome were laid bare in 2005, and the high-resolution (three-Ångström) structure of a eukaryotic ribosome—that of yeast—was published by a French team in December of 2011.[24]

The bacterial ribosome has two major components, called the 30S and the 50S subunits, which drift apart and back together during its operation. The small 30S subunit is the part of the ribosome that reads the genetic code; the larger 50S subunit is where proteins are made. The 30S unit was studied in atomic detail by Yonath and independently by Venkatraman Ramakrishnan, at the Medical Research Council's Laboratory of Molecular Biology, in Cambridge, England. They discovered, for example, an "acceptor site," the part of the 30S subunit that recognizes and monitors the accuracy of the match between messenger and transfer RNAs. Details of the molecular structure reveal how the ribosome enforces the pairing of the first two letters of RNA code: molecules flip out to "feel" for a groove in the double helix of well-matched RNAs to ensure that the code is read with high fidelity. A "wobble" makes this mechanism less stringent in checking the third of the group of three letters that corresponds to a protein building block. This is consistent with the observation that a single transfer RNA—and thus its amino acid—can match up with more than one three-letter code on messenger RNA. For example, the three-letter codes for the amino acid L-phenylalanine are both UUU and UUC.

In complementary work, Harry F. Noller, of the University of California, Santa Cruz (who'd begun his research out of a fascination with the way molecules moved), published the first detailed images of a complete ribosome in 1999, and then in much finer detail in 2001. His work revealed how molecular bridges form and fall during its operation.[25] The ribosome machine contains compression and torsion springs made of RNA to keep the subunits tethered together as they shift and rotate with respect to each other. Its smaller subunit moves along messenger RNA and also binds to transfer RNA, which connects the genetic code on one end with amino acids on the other. The amino acids are linked together into proteins by the larger subunit, which also binds to the transfer RNA. This way, the ribosome is able to ratchet RNAs laden with amino acids through its heart at a rate of fifteen per second and coordinates how they are linked with the growing protein.

Many antibiotics work by disrupting these functions of bacterial ribosomes. Fortunately, though bacterial and human ribosomes are similar, they are sufficiently distinct that antibiotics can bind to and block bacterial ribosomes more effectively than they can human ribosomes. The aminoglycosides tetracycline, chloramphenicol, and erythromycin all work to kill bacterial cells by interfering with the ribosome function.

Yonath, Ramakrishnan, and Thomas A. Steitz would share the 2009 Nobel Prize for Chemistry for their efforts to reveal the workings of this marvelous machine.

As the field of genomics has progressed, RNA has taken on greater importance. According to the central dogma, RNA functioned as a mere middleman, carrying out the commands encoded in DNA. In that model the DNA's double helix unwinds, and its genetic code is copied onto a single-stranded mRNA. In turn, the mRNA shuttles the code from the genome to ribosomes. It was also widely believed that non-protein-coding DNA was "junk DNA." Both perceptions changed in 1998, when Andrew Fire, of the Carnegie Institution for Science, in Washington, D.C., Craig Cameron Mello, of the University of Massachusetts, and colleagues published evidence that double-stranded RNA produced from non-coding DNA can be used to shut down specific genes, which helped explain some puzzling observations, notably in petunias.[26] Now it has become clear that some DNA codes for small RNA molecules that, like switches, play a critical role in how and to what extent genes are used. All the information in a living cell ultimately resides in the precise order of the nucleic acids and amino acids—in DNA, RNA, and proteins. The process of maintaining this extraordinary degree of order in a genome is bound by the sacred laws of thermodynamics. Chemical energy must be burned to enable molecular machines to harness thermal motion. The cell also requires a constant supply of that energy to form the covalent bonds between the subunits as well as to organize these subunits in the correct order, or sequence. At the heart of this storm of chemical turmoil lies a relatively rock-steady set of instructions, those held in the DNA code.

When discussing the genetic code of inheritance, Schrödinger had good reason for envisioning an "aperiodic crystal": he wanted to emphasize the fact that hereditary information is stored, and used the term *crystal* to "account for the permanence of the gene." That is not the case for the protein robots encoded in our genes, which are unstable and quickly break down. With rare exceptions, proteins have a lifetime that ranges from a matter of seconds to days. They have to endure the tumult within a cell, where heat energy sends molecules ricocheting around. Proteins can also misfold into inactive and often toxic aggregates, a process that is central to some well-known diseases.

At any given moment, a human cell typically contains thousands of different proteins, with some being manufactured and others being discarded as needed for the cell's continued well-being. A recent study of one hundred proteins in human cancer cells[27] revealed protein half-lives that ranged between forty-five minutes and twenty-two and a half hours. Cells turn over, too. Every day, five hundred billion blood cells die in an individual human. It is also estimated that half of our cells die during normal organ development. We all shed about five hundred million skin cells every day. As a result, you shed your entire outer layer of skin every two to four weeks. That's the dust that accumulates in your home; that's you. If you're not constantly synthesizing new proteins and cells, you die. Life is a process of dynamic renewal. Without our DNA, without the software of life, cells perish very rapidly, and thus so does the organism.

That the linear chains of amino acids defined by the genetic code fold up into the proper shapes to carry out their particular functions seems, at first sight, little short of miraculous. Not all of the rules that guide protein-folding are yet understood, which is not surprising, given that there are millions to trillions of possible folding configurations for a typical chain of amino acids, or polypeptide. In order to calculate all of the possible conformations of a protein to a predicted thermodynamically stable state, California's Lawrence Livermore National Laboratory joined forces with IBM to spawn Blue Gene, a line of supercomputers that can carry out a trillion or so floating-point operations per second (that is, one petaFLOPS).

A protein with one hundred amino acids can fold in myriad ways, such that the number of alternate structures ranges from 2^{100} to 10^{100} possible conformations. For each protein to try every possible conformation would require on the order of ten billion years. But built into the linear protein code are the folding instructions, which are in turn determined by the linear genetic code. As a result, with the help of Brownian motion, the incessant molecular movement caused by heat energy, these processes happen very quickly—in a few thousandths of a second. They are driven by the fact that a correctly folded protein has the lowest possible free energy, so that, like water flowing to the lowest point, the protein naturally achieves its favored shape.

The correctly folded conformation that ensures that the enzyme can work properly involves moving from a high degree of entropy and free energy to the thermodynamically stable state of decreased entropy and free energy. This process can actually be viewed for a protein called villin, thanks to a computer simulation.[28] Spreading the action of six-millionths of a second out over several seconds, the simulation shows how heat energy makes the initial linear chain of eighty-seven amino acids jiggle; the linear protein shivers this way and that and, over the course of just six microseconds, goes through many different conformations on its way to the final fold. Imagine how much evolutionary selection went into this jittery dance, given that the protein's amino-acid sequence determines not only its rate of folding but its final structure—and, hence, its function.

The competition between productive protein-folding and potentially harmful versions led to the early evolution of cellular protein "quality control," in the form of another group of specialized molecular machines. These "molecular chaperones" aid protein-folding and block the formation of harmful aggregates, as well as dismantle aggregates that do form. So, for example, the Hsp70 and Hsp100 chaperones disassemble aggregates, while Hsp60 consists of various proteins that form a kind of barrel with a cap, so that, when inside, an unfolded protein can achieve the right shape. Not surprisingly, chaperone malfunction underlies a range of neurodegenerative diseases and cancers.

The most common single-gene-hereditary disease in Caucasians—affecting around one in 3,500 births in the United States—is cystic fibrosis, an example of a misfolding, misbehaving protein. It's caused by a defect in the gene that codes for a protein called cystic fibrosis transmembrane conductance regulator (CFTR). This protein regulates the transport of the chloride ion across the cell membrane; when it's faulty, a wide range of symptoms results. As one example, the imbalance of salt and water in patients with cystic fibrosis makes their lungs clog up with sticky mucus, which provides a growth matrix for disease-causing bacteria. Lung damage caused by repeated infections is the leading cause of death for people with the disease. Recently, scientists have shown[29] that by far the most common mutation underlying cystic fibrosis hinders the dissociation of the transport-regulator protein from one of its chaperones. As a result, the final steps in normal folding cannot occur, and normal amounts of active protein are not produced.

Degradation of protein aggregates and protein fragments is of vital importance, because they can form build-ups, or plaques, that are highly toxic. When garbage removal halts as a result of a strike, and malodorous detritus piles up on the streets, traffic slows and the risk of disease rises, and a city rapidly becomes dysfunctional. The same is true of cells and organs. Alzheimer's disease, the tremor of Parkinson's, and the relentless decline caused by Creutzfeldt-Jakob disease (the human form of mad cow disease) all result from the accumulation of toxic, insoluble protein aggregates.

A number of protein machines are designed to cope with mistakes in protein synthesis and folding. The proteasome is responsible for the elimination of abnormal proteins by proteolysis, a peptide-bond-breaking reaction carried out by enzymes called proteases. This particular machine consists of a cylindrical complex containing a "core" of four rings, stacked like bagels, each made of seven proteins. Within the central core, target proteins are marked for degradation with ubiquitin molecules, small proteins that are present throughout the cell. Around three decades ago, this basic mechanism of cellular waste disposal was elucidated by three scientists, Aaron Ciechanover, Avram Hershko, and Irwin A. Rose, for which they won the Nobel Prize in Chemistry in 2004.

The life span of every protein robot in the cell is preprogrammed in the genetic code. The effect of this program varies slightly according to the branch of life. For example, both *E. coli* and yeast cells contain the enzyme beta-galactosidase, which helps break down complex sugars; however, the half-life of this enzyme is highly dependent on the amino acid at the end of the protein (the N-terminal amino acid). When there is an arginine, lysine, or tryptophan as the N-terminal amino acid in beta-galactosidase, the protein half-life is 120 seconds in *E. coli* and 180 seconds in yeast. With serine, valine, or a methionine as the N-terminal amino acid there is a significant extension in half-life, to more than ten hours in *E. coli* and more than thirty hours in yeast. This is what is called the N-end rule[30] pathway of protein degradation.

Protein instability and turnover illustrate that cellular life itself would be very short if cells were only membrane sacs—vesicles—containing proteins but no genetic programing. All cells will die if they cannot make new proteins on a continuous basis to replace those that are damaged or misfolded. In an hour or even less a bacterial cell has to remake of all its proteins or perish. The same is true for cell structures, such as the cell membrane: the turnover of the phospholipid molecules and membrane transporters is such that, if they were not continually replenished, the membrane would break down and spill the cell's contents. When culturing cells in the lab, a simple test for viable candidates is to determine whether the membrane is leaky enough to allow a large dye to enter. If the dye is able to penetrate the cells, they are clearly dead.

There is also protein machinery that degrades and destroys old or failing cells in multicellular organisms. This process of programmed cell death, known as apoptosis, is a crucial part of life and development. Of course, dismantling something as complex as a cell requires an exquisite feat of coordination. The apoptosome, a protein complex nicknamed "the seven-spoked death machine," uses a cascade of caspases—protein-digesting enzymes, or proteases—to initiate destruction. These caspases are responsible for dismantling key cellular proteins, such as cytoskeletal proteins, which leads to the typical changes observed in the shape of cells undergoing apoptosis. Another hallmark of apoptosis is the fragmentation of the DNA software. The caspases play an important role in this process by activating an enzyme that cleaves DNA, DNase. As a result, they inhibit DNA repair enzymes, allowing the breakdown of structural proteins in the nucleus of the cell.

We might think of our bodies as a pattern of proteins in space, but due to the constant turnover of their components, this pattern is a dynamic one. Schrödinger grasped this when he spoke of "an organism's astonishing gift of concentrating a 'stream of order' on itself and thus escaping the decay into atomic chaos—of 'drinking orderliness' from a suitable environment."

Finally, we should consider what ultimately drives all the frantic activity and turnover within each and every cell. If there was a candidate for a vital force to animate life, it is the one that first entranced Robert Brown (1773–1858) in 1827, when the Scottish botanist became fascinated by the incessant zigzag motion of fragments in pollen grains, a phenomenon that would come to be named after him (unless you are French, that is—they argue that similar observations were reported in 1828 by botanist Adolphe-Théodore Brongniart, 1801–1876). What puzzled Brown was that this microscopic motion did not arise from currents in the fluid, or from evaporation, or from any other obvious cause. At first he thought that he had glimpsed "the secret of life," but after observing the same kind of motion in mineral grains he discarded that belief.

The first key step in our current understanding of what Brown had witnessed came more than seventy-five years after his discoveries, when Albert Einstein [1879–1955] demonstrated how the tiny particles were being shoved about by the invisible molecules that made up the water around them. Until Einstein's 1905 paper, a minority of physicists (notably Ernst Mach [1838–1916]) still doubted the physical reality of atoms and molecules. Einstein's notion was eventually confirmed with careful experiments conducted in Paris by Jean Baptiste Perrin (1870–1942), who was rewarded for this and other work with the Nobel Prize in Physics in 1926.

Brownian motion has profound consequences when it comes to understanding the workings of living cells. Many of the vital components of a cell, such as DNA, are larger than individual atoms but still small enough to be jostled by the constant pounding of the surrounding sea of atoms and molecules. So while DNA is indeed shaped like a double helix, it is a writhing, twisting, spinning helix as a result of the forces of random Brownian motion. The protein robots of living cells are only able to fold into their proper shapes because their components are mobile chains, sheets, and helices that are constantly buffeted within the cell's protective membrane. Life is driven by Brownian motion, from the kinesin protein trucks that pull tiny sacks of chemicals along microtubules to the spinning ATP synthase.[31] Critically, the amount of Brownian motion depends on temperature: too low and there is not enough motion; too high and all structures become randomized by the violent motion. Thus life can only exist in a narrow temperature range.

Within this range, the equivalent of a Richter 9 earthquake rages continuously inside cells. "You would not need to even pedal your bicycle: you would simply attach a ratchet to the wheel preventing it from going backwards and shake yourselves forward," according to George Oster and Hongyun Wang, of the Department of Molecular and Cellular Biology at the University of California, Berkeley.[32] Protein robots accomplish a comparable feat by using ratchets and power strokes to harness the power of Brownian motion. Due to the incessant random movement and vibrations of molecules, diffusion is very rapid over short distances, which enables biological reactions to occur with tiny quantities of reactants in the extremely confined volumes of most cells.

Now that we know that the linear code of DNA determines the structure of the protein robots and RNAs that run our cells and, in turn, that the structure determines the functions of the protein and RNAs, the next question is obvious: how do we read and make sense of that code so that we can understand the software of life?

4 Digitizing Life

The early days of molecular biology were marked by what seemed to many to be an arrogant cleavage of the new science from biochemistry. However, our argument was not concerned with the methods of biochemistry, but only their blindness in ignoring the new field of the chemistry of information.

—Sydney Brenner, 2005[1]

This is now the era of digital biology, in which the proteins and other interacting molecules in a cell can be viewed as its hardware and the information encoded in its DNA as its software. All the information needed to make a living, self-replicating cell is locked up within the spirals of its double helix. As we read and interpret that code, we should, in the fullness of time, be able to completely understand how cells work, then change and improve them by writing new cellular software. But, of course, that is much easier to say than to do in practice: studies of this DNA software reveal it to be much more complex than we had thought even a decade ago.

While the first linear amino acid sequence of a protein (insulin) was determined in 1949 by Fred Sanger, the processes for reading DNA took longer to develop. In the 1960s and 1970s, progress was slow, and sequencing was measured in terms of a few base pairs per month or even per year. For example, in 1973 Allan Maxam and Walter Gilbert, of Harvard University, published a paper describing how twenty-four base pairs had been determined with their new sequencing method.[2] Meanwhile, RNA sequencing was also underway and progressed a bit faster. Still, compared with the abilities of today's technology, the effort required to read even a few letters of code was heroic.

Most people are aware of genomics from the first human-genome decoding, which culminated in my appearance at the White House in 2000, along with my colleague-competitors and President Clinton, to unveil the human genome sequence. In fact, the first ideas about decoding DNA date back more than half a century, to when the atomic structure of DNA was proposed by Watson and Crick. A major leap in our knowledge occurred when, in 1965, a group led by Robert Holley, from Cornell University, published the sequence of the seventy-seven ribonucleotides of alanine transfer RNA (tRNA) from the yeast cell *Saccharomyces cerevisiae*,[3] as part of the effort to work out how tRNAs helped combine amino acids into proteins. RNA sequencing continued to lead the way when, in 1967, Fred Sanger's group determined the nucleotide sequence of the 5S ribosomal RNA from *E. coli*, a small RNA of 120 nucleotides.[4] The first actual genome that was successfully decoded was an RNA viral genome: the bacteriophage MS2 was sequenced in 1976 by the laboratory of Walter Fiers, at the University of Ghent, in Belgium. Fiers had studied bacteriophages (which hijack bacterial cells to reproduce) with Robert L. Sinsheimer, at the California Institute of Technology (Caltech), and then with Har Gobind Khorana, in Madison, Wisconsin.

The DNA-sequencing technology that made it possible for me to sequence the human genome originated in the mid-1970s, when Fred Sanger's team in Cambridge developed new DNA-sequencing techniques, the first being "plus-minus" sequencing, followed by what Sanger named dideoxy DNA sequencing but which in his honor is now called Sanger sequencing. Sanger sequencing uses dideoxynucleotides or terminator nucleotides to stop the DNA polymerase from adding additional nucleotides to the growing DNA chain; dideoxynucleotides lack a hydroxyl (-OH) group, which means that, after being linked by a DNA polymerase to a growing nucleotide chain, no further nucleotides can be added. By attaching a radioactive phosphate to one of the four nucleotides, to label the fragments, it was possible to read the order of the As, Ts, Cs, and Gs by exposing the gel used to separate one base from another to X-ray film.

Sanger's team used his new sequencing tools to determine the first DNA viral-genome sequence, that of the bacteriophage phi X 174,[5] which was published in *Nature* in 1977. Clyde Hutchison (now with the Venter Institute) was a visiting scientist in Sanger's lab (from University of North Carolina where he was a faculty member since 1968) and contributed to the sequencing of the phi X 174 genome. In the 1950s Sinsheimer, using light scattering, had estimated the phi X 174 genome size to be around 5,400 bases and was gratified when Sanger revealed that the actual number was 5,386.[6]

I had completed my Ph.D. at the University of California, San Diego (UCSD), two years before Sanger's article appeared and had since moved to the State University of New York at Buffalo to start my independent research and teaching career. I missed the Sanger publication at the time because it was the middle of the deadly blizzard of '77, and my son was born two weeks after the publication.[7] My lab at the time was working on the isolation and characterization of the proteins at the site where signals are passed between nerve cells, called neurotransmitter receptors.

DNA sequencing progressed gradually over the decade following the work on the phi X 174 genome. While Sanger sequencing became the world standard, it was slow, very laborious, and required use of substantial quantities of radioactive phosphorus, which had a half-life of only a couple of weeks. Also, reading sequencing gels was more of an art than a science. In his second Nobel Prize lecture, Sanger described the tedious effort involved in early DNA sequencing, concluding, "[It] seemed that to be able to sequence genetic material a new approach was desirable."[8]

In 1984 I had moved my research team to the National Institutes of Health, and we began teaching ourselves molecular biology with the help of some good molecular-biology cookbooks and my interactions with Marshall Nirenberg and his lab. During my first year at NIH we sequenced only one gene, the human-brain adrenaline receptor,[9] using the radioactive Sanger sequencing, but it took the better part of a year. Like Sanger, I was certain that there had to be a better way. Fortunately, it was around this time that Leroy Hood and his team at Caltech published a key paper describing how they replaced the radioactive phosphate with four different fluorescent dyes on the terminator bases of DNA, which, when activated with a laser beam, could be read sequentially into a computer.[10] I obtained one of the first automated DNA-sequencing machines from the new company Applied Biosystems just as serious discussions got under way about a wild proposal to sequence the entire human genome.

Using the new DNA-sequencing technology coupled with computer analysis, my lab rapidly sequenced thousands of human genes by a new method I had developed which focused on relatively short sequences which my team had named expressed sequence tags (ESTs).[11] The EST method involved sequencing the expressed genetic material, messenger RNA (after converting it into complementary DNA). Although we successfully discovered several thousand human genes with the EST method, my approach was not immediately appreciated by many who saw it as a threat to the traditional way of doing gene discovery, because we could discover more new genes per day than the entire scientific community had over the previous decade. The situation was not helped when the U.S. government decided to file patents on all the genes identified by my team. While our discoveries provoked attacks and controversy, they also resulted in some attractive offers, including one to form my own basic science research institute, which I accepted in 1992. I named it The Institute for Genomic Research (TIGR), and it was there, in Rockville, Maryland, that we built the world's largest DNA-sequencing factory, using the latest versions of the automated DNA-sequencing machines.

The course of the history of genomics changed in 1993, after a chance encounter at a scientific meeting in Bilbao, Spain, where I had outlined our rapid advances in discovering genes. Many in the audience appeared to be shocked by the voluminous results of our EST effort and by the nature of our discoveries—notably the genes responsible for non-polyposis colon cancer, discovered in collaboration with Bert Vogelstein of the Johns Hopkins Kimmel Cancer Center, Baltimore.

Once the crowd that had come up to ask direct questions had dissipated, I was confronted by a tall, kindly-looking man with silver hair and glasses. "I thought you were supposed to have horns," he said, referring to the demonic image that the press had often used to portray me. He introduced himself as Hamilton Smith, from Johns Hopkins. I already knew of Ham through his huge reputation in the field and his Nobel Prize, and I took an instant liking to him—he had clearly decided that he was going to make up his own mind about me and my science and not have his opinion dictated by others.[12]

Ham had by then had a long, productive career and, at 62, had been thinking of retiring. As we talked at the bar and then dinner following my lecture, however, he made an interesting suggestion: he proposed that his favorite bacterium, *Haemophilus influenzae*, from which he had isolated the first restriction enzymes, would be an ideal candidate for genome sequencing using my approaches.

Our first joint project got off to a slow start, as Ham explained that there were problems with producing the libraries of clones containing *H. influenzae* genome fragments. Only years later did he reveal that his colleagues at Johns Hopkins had been less than impressed with our project, viewing me with suspicion because of the EST furor and fearful that his association with me would ruin his reputation. Even though many of them would spend their careers studying *H. influenzae*, they did not immediately see the value of obtaining its entire genome sequence. Ham was eventually forced to sidestep his own team, as I had also done years earlier in my work with ESTs.[13]

Ham began to collaborate with me at TIGR. Our work on the project began in 1994 and involved most of my scientific team. Unlike Sanger's lab years earlier with phi X 174, which used isolated unique restriction fragments for sequencing one at a time, we relied completely on randomness. We broke up the genome into fragments in a mixed library and randomly selected twenty-five thousand fragments to obtain sequence reads of around five hundred letters each. Using a new algorithm developed by Granger Sutton, we began to solve the greatest biological puzzle to date, reassembling those pieces into the original genome. In the process we developed a number of new methods to finish the genome. Every single one of the base pairs of the genome was accurately sequenced and the twenty-five thousand fragments accurately assembled. The result was that the 1.8 million base pairs of the genome were re-created in the computer in the correct order.

The next step was to interpret the genome and identify all its component genes. As the first to examine the gene complement of a living self-replicating organism, I wanted to do much more than simply report the sequence. The team spent substantial time working out what the gene set said about the life of the organism. What did the software that programmed the structures and functions of life *mean*? We wrote up our results in a scientific paper that was rapidly accepted for publication in the journal *Science* and was scheduled to appear in June 1995. Rumors of our success were circulating weeks beforehand. As a result, I was invited to deliver the president's lecture at the annual meeting of the American Society of Microbiology, which was being held in Washington, D.C., on May 24, 1995, and I accepted with the understanding that Ham would join me on stage. The pressure really came to bear on me when the society's president, David Schlessinger, of Washington University, in St. Louis, announced what he described as an "historic event."

With *Haemophilus influenzae* we had transformed the double helix of biology into the digital world of the computer, but the fun was only now beginning. While we had used its genome to explore the biology of this bacterium and how it causes meningitis and other infections, we had in fact sequenced a second genome to validate the method: the smallest one known, that of *Mycoplasma genitalium*. When I ended my speech, the audience rose in unison and gave me a long and sincere ovation. I had never before seen so big and spontaneous a reaction at a scientific meeting.[14]

That was a very sweet moment. My team had become the first ever to sequence the genetic code of a living organism, and of equal significance was the fact that we had done so by developing a new method, which we named "whole genome shotgun sequencing." This feat marked the start of a new era, when the DNA of living things could be routinely read so that they could be analyzed, compared, and understood.

After we had finished the *Haemophilus influenzae* genome, I wanted to sequence a second genome so that we would be able to compare two genomes to aid in understanding the basic set of genes required for life. At that time Clyde Hutchison, at the University of North Carolina, Chapel Hill, had come up with an attractive candidate with the smallest known genome size: a species of *Mycoplasma genitalium*, with fewer than five hundred genes. It seemed that this genome would complement our work on *H. influenzae*, because it came from a different group of bacteria. Gram staining, so named after its inventor, Hans Christian Gram (1853–1938), categorizes all species of bacteria into two groups, depending on how they react to a stain: Gram-positive (such as *Bacillus subtilis*, for example) results in a purple/blue color, while Gram-negative organisms (such as *H. influenzae*) result in a pink/red color. *M. genitalium* is thought to be evolutionarily derived from a bacillus species and is thus classified as a member of the Gram-positive bacteria.

The genome sequencing required only three months to complete, and in 1995 we published the 580,000 base-pair genome of *Mycoplasma genitalium* in *Science*.[15] While our accomplishment would ultimately serve as the foundation for the quest to create a synthetic cell, it had more immediate implications. In its aftermath we were able to launch a new discipline, known as comparative genomics. By comparing the first two genomes sequenced in history, we could look for common elements associated with a living self-replicating life form. Comparative genomics exploits one of the most exciting findings of biology: when evolution yields a protein structure that performs a critical biological function, evolution tends to use the same structure/sequence over and over again.

The genes that control the fundamental process of cell division in yeast, for example, are similar to the ones our own cells use.[16] Because the gene that codes for DNA polymerase had been identified, sequenced, and functionally characterized from the bacterium *E. coli*, our team could use this information to search for similar sequences in the putative gene sequences from *H. influenzae*. If any of the DNA sequences was a close match to that of the *E. coli* DNA polymerase gene, we could infer that the *H. influenzae* gene was likewise a DNA polymerase. The problem was that in 1995 the gene databases were sparsely populated, so there was not much with which to compare our genome. As a result almost 40 percent of the putative genes in our sequenced genomes had no matches in the database.

Our *Science* paper on *M. genitalium* described how we used the data from both sequenced genomes to ask basic questions about the recipe of life: what were the key *differences* in the gene content of the two species? There are about 1,740 proteins in *H. influenzae*, each coded for by a specific gene, and another eighty genes code for RNAs. *M. genitalium* has only 482 protein-coding genes and forty-two RNA genes. The *M. genitalium* genome is smaller in part because it lacks all the genes to make its own amino acids (it is able to acquire them from its human host). Like *M. genitalium* we also have "essential amino acids," such as valine and tryptophan, which our cells cannot make but have to obtain from our diet.

Perhaps an even more interesting question is what genes do these quite different microorganisms *share*? If the same genes are found to be present in many different types of organisms, they take on a much greater significance. Common genes suggest a common ancestor and that they could in fact be central to the very process of life itself. A key paragraph from our 1995 paper reads, "A survey of the genes and their organization in *M. genitalium* permits the description of a minimal set of genes required for survival."

We began to think about the basic gene set of life. What is the smallest number of genes that is required for a cell to survive and thrive? We hoped that the genes held in common by these bacteria from two different groups would provide a glimpse of the critical gene set.

One reflection of the poor state of our biological knowledge in 1995 was the fact that we had no idea of the function of 736 genes, or 43 percent, of the total in *H. influenzae* and 152 genes, or 32 percent, of the *M. genitalium* genes. As the papers were being written we had many discussions about life and whether *M. genitalium* actually represented a true minimal gene set. The *M. genitalium* paper itself alluded to our discussions when it concluded, "Comparison of [newly sequenced genomes] with the genome sequence of *M. genitalium* should allow a more precise definition of the fundamental gene complement for a self-replicating organism and a more comprehensive understanding of the diversity of life." Other groups also began work on our data from the first two published genomes. Eugene Koonin, at the NIH, hailed this development as marking a new era in genome science and concluded from a computational study that there was very little gene diversity in microbes, based on the similarity between the gene sets of a Gram-negative (*H. influenzae*) and a Gram-positive bacteria (*M. genitalium*).[17] However, our next genome project would, at a stroke, change the worldview of gene diversity.

In 1996 we purposely chose an unusual species for our third genome effort: *Methanococcus jannaschii*. This single-cell organism lives in an extraordinary environment, a hydrothermal vent where hot, mineral-rich liquid billows out of the deep seabed. In these hellish conditions the cells withstand over 245 atmospheres—equivalent to the crushing pressure of 3,700 pounds per square inch—and temperatures of around eighty-five degrees centigrade (185 degrees Fahrenheit). That in itself is remarkable, as most proteins denature at around fifty to sixty degrees centigrade, which is why egg white becomes opaque when cooked. Unlike life on the surface of Earth, which is dependent on sunlight, *Methanococcus* is an autotroph, meaning that it makes everything it needs for its sustenance from inorganic substances. Carbon dioxide is the carbon source for every protein and lipid in a *Methanococcus* cell, which also generates its cellular energy by converting carbon dioxide into methane. *Methanococcus* is from the proposed third branch of life, called the Archaea, discovered by Carl Woese, of the University of Illinois, Urbana, in 1977.[18] The *Methanococcus* genome was chosen in collaboration with Woese as the first of the Archaea to be sequenced and analyzed.

The sequence did not disappoint. The *Methanococcus* genome[19] broadened our view of biology and the gene pool of our planet. Almost 60 percent of the *Methanococcus* genes were new to science and of unknown function; only 44 percent of the genes resembled anything that had been previously characterized. Some of *Methanococcus*'s genes, including those associated with basic energy metabolism, did resemble those from the bacterial branch of life. However, in stark contrast, many of its genes, including those associated with information processing, and with gene and chromosome replication, had their best matches with eukaryote genes, including some from humans and yeast. Our genome study appeared on the front page of every major paper in America and made headlines in much of the rest of the world: *The Economist* settled on "Hot Stuff," while *Popular Mechanics* announced "Alien Life on Earth," a theme also pursued by the *San Jose Mercury News* with "Something Out of Science Fiction."[20] Recent studies suggest that eukaryotes are a branch of the Archaea, which if true would again return us to two major branches of life.[21]

That same year, 1996, NASA made headlines around the world when it published what some thought to be evidence of microbial life on Mars. Everett Gibson and his colleagues at the agency announced that they had discovered fossils a few tens of nanometers across in a meteorite known as ALH 84001. This was a sensational find, because ALH 84001 had been blown out of the surface of the Red Planet and had then fallen to Earth roughly thirteen thousand years ago.

This news of microbial Martians, accompanied by intriguing images of tiny blobs and microscopic sausages, stimulated even more discussions as to what might constitute a minimal genome. With a simple back-of-the-envelope check we worked out the volume of the reported "nanobacterium," which proved to be so small that it could not possibly contain any DNA or RNA molecules. It is now clear that the structures seen in ALH 84001 are not from living things and that crystal growth mechanisms are able to produce deposits that resemble primitive cells.[22]

Over the next few years my team would go on to sequence a large number of unusual species genomes, including one that was inspired by the pioneering work of Barry Marshall, in Australia. He and pathologist Robin Warren believed that spiral-shaped bacteria, later named *Helicobacter pylori*, were responsible for stomach ulcers. I had been inspired by how Marshall had persevered, despite his work's being constantly challenged. His peers did not want to believe that bacteria, and not stress, could be the cause of ulcers. In 1984 Marshall had had the courage of his convictions to swallow a solution of the bacteria. He soon threw up and developed stomach inflammation. Eventually, his persistence paid off. His research made it possible for millions of people to be treated with antibiotics, which also reduced their risk of developing gastric cancer, instead of having to take daily acid-reducing drugs. We published the *Helicobacter pylori* genome in 1997,[23] and in 2005 Marshall was awarded the Nobel Prize in Medicine.[24]

Because single-celled life has been in existence for close to four billion years, it has diversified to occupy a vast range of environments, from the freezing deserts of Antarctica to hot acidic springs. This ability to live at the extremes has earned these organisms that live in marginal environments the name "extremophiles." By probing life at the limits, as we had already done with *Methanococcus*, we thought we could gain the most from comparative genomics. The next extremophile genome we sequenced was *Archaeoglobus*, which lives in oil deposits and hot springs. The organism uses sulfate as its energy source but can eat almost anything.[25] Our first analysis of more than two million letters of its genome revealed that one quarter of its genes were of unknown function (two thirds of these mystery genes are shared with *M. jannaschii*) and another quarter encoded new proteins.

Our sequencing of the first two bacterial genomes and the first Archaea genome, and the publication of the yeast genome[26] by a large consortium of labs, gave the world the first views of the genomes of all three branches of life. What did these data tell us about the basic recipe of life? Our attempts to identify life's elemental genes drove us down several experimental paths. Our plan from the beginning was to approach the goal of understanding a minimal self-replicating life form from a number of directions. While synthesizing the genome would be the ultimate solution, we needed a great deal of information about basic cellular life that was lacking in the scientific literature.

The most obvious approach was to knock out genes in the *M. genitalium* genome to try to establish which are essential: remove or disable a gene and, if the organism continues to live, you can assume that particular gene did not have a critical role; if the organism dies, the gene was clearly essential. The idea was simple and had been successfully used before, in a range of species. Mario Capecchi, of the University of Utah, Oliver Smithies, at the University of North Carolina, Chapel Hill, and Martin Evans, at Cardiff University, in the United Kingdom, shared the 2007 Nobel Prize for their work in the 1980s on the technology used to create knockouts in mice, where one or more genes have been selectively turned off.

The practical obstacles in applying these methods to *M. genitalium* were another matter, however. Performing gene knockouts in a species like yeast is relatively easy, thanks to the range of genetic tools available for that species. Such tools were completely lacking for the mycoplasmas, as are the means for making multiple serial gene changes.

One of the fundamental molecular-biology tools is antibiotic selection. With antibiotic selection, cells in which gene changes have been made are selected for by killing all the unmodified cells with an antibiotic. The modified cells survive because the DNA plasmids that are used to introduce new genes to them also contain genes that code for enzymes that confer resistance to the antibiotic. While this technique is the basis of most molecular-biology experiments, there are unfortunately only a few antibiotic selection systems available, severely limiting the number of serial gene changes that can be made.

To solve one of the problems, Clyde Hutchison came up with a unique approach that we called "whole genome transposon mutagenesis," in which a small unit of DNA called a transposon disrupts a gene, which allowed us to conclude whether the gene was essential. Transposons are relatively short DNA sequences that contain the necessary genetic elements to enable them to insert themselves either into specific sequences or randomly in a genome. In work that would earn her the Nobel Prize in 1983, American Barbara McClintock had discovered transposable elements in maize, where they altered the patterns of pigmentation of kernels.[27] You can think of transposons as selfish genes, similar to viruses, which "infect" a genome. It turns out that a great portion of your genome is composed of such DNA parasites. They are significant, not least because they can cause genetic diseases if they insert into and disrupt the function of a key gene.

We chose a transposon (Tn4001) isolated from *Staphylococcus aureus* to randomly insert itself in the *M. genitalium* genome to disrupt gene functions. We grew the cells that survived the insertions and isolated and sequenced their DNA, starting with a sequencing primer that bound only to the transposon, to determine precisely where in the genome the transposon had ended up. If the Tn4001 inserted into the middle of a gene and the cells survived, then we scored that gene as nonessential for life.

After bombarding the genome with transposons, we scored all the genes that did not have transposon insertions in living cells to be essential. Once we had completed our analysis of the data, we realized that this absolute scoring system was naïve and that genes and genomes are context-specific and that genes alone cannot define life. Because all cells obtain key nutrients and chemicals from their environment, if the environment changes, then the genes that are required for life in that new environment also need to change.

Membrane transport proteins are responsible for moving essential nutrients from the environment into cells. For example, *M. genitalium* can grow independently on two sugars, glucose and fructose, and there are genes that encode for a specific protein-transporter for each sugar. In our transposon-insertion studies, both genes showed up in the nonessential group, which initially surprised us, because they are central to the way the organism feeds. However, we realized that the media that we used to grow the *M. genitalium* cells contained both glucose and fructose, which meant that, if the gene for either transporter was knocked out, the cell simply switched to consuming the other sugar. In contrast, if we grew the cells on just one sugar, then the cells died when that sugar transporter was knocked out. With some functions, such as sugar metabolism, it is not hard to distinguish the "conditionally essential genes," but for the cellular functions and genes that are unknown, there is not an obvious way to ascertain if another gene is providing backup for the disrupted one.

This point was emphasized when we extended the studies to the related species *Mycoplasma pneumonia*, the closest known relative of *M. genitalium*, which has a genome size of 816,000 base pairs, 236,000 base pairs larger than that of *M. genitalium*. Once again we wanted to use the transposon-insertion studies in conjunction with comparative genomics to determine the minimal number of genes required for life. The *Mycoplasma pneumonia* genome includes genes that evolved from a common ancestral gene (ortholog) of virtually every one of the 480 *M. genitalium* protein-coding genes, plus an additional 197 genes. That raised a tantalizing possibility: could it be that the 480 genes the two species have in common are already close to constituting a minimal gene set? Our starting assumption was that the 197 extra genes in the *M. pneumonia* genome should all be disruptable by transposon insertions, as the very existence of *M. genitalium* would suggest that they are not necessary for life. The results were not very satisfying or informative; we found that 179 genes in *M. pneumonia* were disrupted by transposon insertions, but only 140 of the 197 extra genes were disrupted.

From our combined studies we estimated that in *M. genitalium* 180 to 215 genes are nonessential and that the number of essential genes is

265 to 350. Of the latter, 111 are of unknown function. This was clearly not the precise definition of life that we were seeking. In addition, working through this data, it became increasingly obvious that the genes that were individually dispensable might not be able to all be deleted together.

Given the limits of the molecular-biology tools and the limitations of the transposon data, we concluded that the only way to get to a minimal genome would be to attempt to synthesize the entire bacterial genome from scratch. We would have to chemically synthesize the entire chromosome, using only the essential genes. However, this would be a huge challenge. Even though scientists had been writing small pieces of genetic code for almost half a century, no one had made any DNA constructs that were were even within twenty times the size we needed.

Work on the chemical synthesis of DNA dates back to the 1950s, with the success of Har Gobind Khorana and Marshall Nirenberg, but it was only in the 1980s that substantial progress was made, following the invention of the automated DNA synthesizer by Marvin Caruthers, at the University of Colorado, Boulder. His synthesizer uses four bottles containing the DNA bases A, T, C, and G, and adds one base to another in a prescribed order. In this way, DNA synthesizers can make short stretches of DNA called oligonucleotides. However, the yield and accuracy decrease as the length of the oligonucleotide increases. An entire industry has been built around synthesizing oligonucleotides and shipping them to researchers, because synthetic DNA is used in molecular biology for DNA sequencing and PCR (polymerase chain reactions).

Chemical methods can be used to string together synthetic oligonucleotides to make longer pieces of DNA. When we had first started to discuss synthesizing an entire genome, the largest pieces of DNA that had been made measured only a few thousand base pairs. To build the genome of a viable organism required us to chemically synthesize and assemble almost six hundred thousand base pairs, and as a result we knew that we would need to develop new methods to accomplish this goal. To see if our idea was even remotely possible we decided that we should first attempt a smaller test project. We chose to synthesize the genome of the bacteriophage phi X 174. Aside from being the first DNA virus sequenced, a remarkable and successful attempt to copy the single-stranded genome enzymatically had actually been made by another team more than three decades earlier.

5 Synthetic Phi X 174

It is going to be one of the most important stories that you ever read, your daddy ever read, or your grandpappy ever read. . . . These men have unlocked a fundamental secret of life. It is an awesome accomplishment. It opens a wide door to new discoveries in fighting disease, in building much healthier lives for all human beings. It could be the first step—these great laboratory geniuses say—toward the future control of certain types of cancer.

—President Lyndon Johnson, December 1967[1]

Even though most people have never heard of phi X 174, this simple bacteriophage has already earned its place in history. The phage was the first DNA virus to be sequenced and the first to have its genome artificially copied and activated. Discovered in the sewers of Paris,[2] phi X 174 targets the human-intestine bacterium *Escherichia coli*. One might wonder why so much attention has been focused on a virus seemingly so unremarkable, but the reason is simple: there's not much to this virus when you examine it at the molecular level.

Phi X 174 consists of a circular DNA chromosome—only eleven genes in all—wrapped in an icosahedral "coat" of proteins, including a dozen pentagonal "spikes." Though under an electron microscope the phage looks as beautiful as a flower, in reality it is a cold, geometric form. The virus is no more alive than a crystal of salt. Its life cycle, such as it is, goes as follows: the phage injects its DNA through its spikes into a bacterial cell, where it hijacks the cell's biochemical machinery to create many new viruses. The progeny then erupt from the cell so they can go on to infect even more *E. coli* bacteria.

Before the emergence of DNA sequencing and before the structure of the phage genome was even known, the virus had been re-created in the laboratory in the 1960s by a team at Stanford University, led by the biochemist Arthur Kornberg. The key to this achievement was the discovery by Kornberg of DNA polymerase, which is the pivotal enzyme for DNA replication. Kornberg's lab began by using the newly discovered enzyme to copy DNA in vitro. According to Kornberg's papers, his team initially attempted to copy a bacterial genome but was unsuccessful, due to the inability of the polymerase to read nonstop through the entire genome, which consists of millions of bases of DNA.

After this failed experiment, Kornberg decided to do what my own team was to do thirty years later: he chose a DNA target less ambitious to replicate, that of phi X 174. As one of the early pioneers of gene synthesis and sequencing, Robert Sinsheimer had by then discovered some important details of the phage's life cycle. Though the phi X 174 virus's DNA consists of a single circular strand of DNA, Sinsheimer found that immediately after it had infected its host, enzymes present in the bacterium converted the circle of DNA to the familiar linear double helix. This discovery illuminated a problem encountered by Kornberg in his first efforts to make a copy of the virus genome: while DNA polymerase could copy the entire phi X 174 genome (5,386 base pairs) in a linear form, it could not create the infectious circular form. We all know how to turn a line into a circle, but doing so on the molecular level was not so easy for scientists half a century ago.

Nature had mastered the feat, of course, and several groups of scientists, including Kornberg's, searched for a bacterial enzyme that could join up the two ends of the linear double-stranded DNA to turn it full circle, like a molecular Ouroboros. That search ended in 1967, when five groups discovered DNA ligase, an enzyme capable of linking DNA into a ring. By the end of that year Kornberg had used the DNA ligase to join the ends of the phi X 174 DNA that had been copied using the DNA polymerase. The duplicated DNA was now capable of infecting a bacterium. In this way Kornberg had succeeded in copying the genome of phi X 174, circularizing it, and using the enzymatically copied DNA to infect *E. coli* and produce multiple copies of the virus.

Although he knew he had achieved this feat in broad outline, he did not know what the phi X 174 genome consisted of in terms of its DNA sequence. Kornberg would only realize what he had "brought to life" ten years later, in 1977, when Fred Sanger's team used DNA polymerase in their new sequencing method, which they applied to the phi X 174 genome. Nonetheless, Kornberg's work caused a sensation. On December 14, 1967, Stanford University arranged a press conference for Kornberg to coincide with the publication of his paper in the journal *Proceedings of the National Academy of Sciences*. They had asked journalists in advance not to characterize his achievement as "synthesizing life," as viruses are not living things; they depend on other life to multiply. But they had not briefed everyone.

President Lyndon B. Johnson was scheduled to speak that same day at the Smithsonian Institution to mark the two hundredth anniversary of the *Encyclopædia Britannica*,[3] and his speechwriter had asked Stanford for a paragraph on the DNA work. This was supplied, but as Johnson began to read the prepared statement, he abruptly put it aside and could not contain his excitement as he told his audience about the news that was about to make headlines:

> It is going to be one of the most important stories that you ever read, your daddy ever read, or your grandpappy ever read. . . . These men have unlocked a fundamental secret of life. It is an awesome accomplishment. It opens a wide door to new discoveries in fighting disease, in building much healthier lives for all human beings. It could be the first step—these great laboratory geniuses say—toward the future control of certain types of cancer.

The president also mulled over the issue of a government's "ordaining life," when a state possessed powers that were once thought only the province of nature, or even a god: "This is going to be one of the great problems—one of the big decisions. If you think about some of these decisions the present president is making—it is going to be a kindergarten class compared to the decisions some future president is going to have to make." After this speech, it was hardly surprising that headlines appeared around the world that heralded the birth of the first synthetic life.[4]

I was fortunate to learn about the "secret of life" quote decades later, when I was doing a National Public Radio interview with science correspondent Joe Palca following our announcement of the first synthetic cell, in 2010. (What the president had hailed as the most important story for generations had passed me by when the news first broke because I was serving as a Navy corpsman in Da Nang, Vietnam.) I thought the anecdote was a delightful find and a good illustration of the continuity of scientific thinking, in this case the goal to ultimately understand life by re-creating it, and the ever-present difficulty of how to convey the genuine excitement of science to the public without overdoing it and resorting to hype. I had once met Kornberg through my Ph.D. mentor, Nathan ("Nate") O. Kaplan. I wonder what Kornberg would have made of the remarkable developments in genomics that came in the wake of his early experiments.

The efforts to re-create DNA viruses using the original viral genome as a template would later extend to more primitive viruses with an RNA code, such as poliovirus, and the retroviruses, such as HIV, which also have an RNA code but are duplicated in a host cell using an enzyme that converts RNA into DNA. The 1970 discovery of that enzyme, reverse transcriptase, came as a result of research on RNA tumor viruses and dramatically challenged the central dogma that "DNA makes RNA makes protein." For their independent discovery of reverse transcriptase, Howard Martin Temin, at the University of Wisconsin, Madison, and David Baltimore, of the Massachusetts Institute of Technology, shared the 1975 Nobel Prize in Physiology or Medicine.

Among the RNA viruses are bacteriophages; one of them, the Qbeta phage, was the first to be re-created by reverse transcriptase in the laboratory. This effort was the culmination of remarkable work by the Swiss molecular biologist Charles Weissmann, who is perhaps best known for his research at the University of Zürich on prions and for manufacturing the protein interferon by using recombinant-DNA technology in 1980, a few years after the foundation of the first biotech company, Genentech.[5] Weissmann's earlier pioneering RNA-sequencing efforts with Martin Billeter, in 1969,[6] would inspire Fred Sanger.[7] Then, working with Richard Flavell, Weissmann would, in 1974, open the door to what became known as "reverse genetics," in which the effects on an organism of altering its genetic code are studied, as opposed to classical genetics, where a mutant organism is identified and the responsible DNA mutation is then tracked down.

The advance in the ability to copy RNA viruses also came in 1974, when Weissmann, joined by Tadatsugu Taniguchi, succeeded in generating a complementary double-stranded DNA copy (cDNA) of Qbeta RNA, which they integrated within a plasmid vector. To Weissmann's "surprise and delight," when implanted into *E. coli,* the plasmid gave rise to infectious Qbeta phage. This was the first time such a feat had been achieved.[8] The ability to make viral cDNA would allow genetic manipulations that could not be carried out at the RNA level, so that recombinant DNA technology could be applied to RNA viruses. A few years later, Vincent Racaniello and David Baltimore repeated this experiment at MIT using purified poliovirus RNA and human cancer cells: they too obtained authentic virus particles.[9] Since this work, the genetic codes of members of nearly every virus family have been reported, including hepatitis C,[10] rabies, respiratory syncytial virus, influenza A, measles, ebola, bunyavirus, and the influenza virus responsible for the 1918 pandemic.[11]

Reverse transcriptase and DNA polymerase also contributed to improving the methods used to read the genetic code. Reverse transcriptase is routinely used to create cDNA clones from mRNA, which allows for DNA sequencing of expressed genes. This is the method that I used with my expressed sequence tag (EST) approach. From Sanger's phi X 174 to *H. influenzae* to the human genome, DNA polymerases have played a critical role in DNA sequencing. To read the three billion bases of the human genome, my team developed the whole genome shotgun method, which relied on breaking genomes down into small pieces that could be easily read by the DNA-sequencing machines. In 1999 it required nine months to read and assemble twenty-five million individual sequence-reads into a complete human genome. Today DNA sequencing has advanced to a remarkable degree, as a host of new technologies have emerged that enable single benchtop units to sequence a human genome in one day.

After finishing sequencing the human genome at the company Celera, my own efforts shifted back to investigating the elements required for minimal life and further work in synthetic genomics. Ham Smith left Celera to join me, and to fund this effort we put together a grant application to the Department of Energy. The DOE ran a program called Genomes to Life, which had evolved from its human-genome effort. Early on, the DOE funded my team's sequencing of some of the first genomes to be completed, including *M. genitalium* and *Methanococcus jannaschii*. Eventually the DOE would provide us with $5 million a year for a five-year period—a great start that would make it possible for us to explore two new areas that we believed presented exciting opportunities.

The first was a radical extension of my earlier work on sequencing genomes, which would come to be known as metagenomics. Rather than focusing on one particular species, we would seek to obtain a genetic snapshot of the entire microbial diversity in an environment such as the ocean or the human gut. Many of my peers were doubtful that shotgun sequencing all the organisms in a sample of seawater would work, because we were dealing with a soup containing vast numbers of different species. The skepticism arose because the complex process would involve simultaneously sequencing the thousands of genomes present in a sample and then using the computer to reconnect and assemble only the correct fragments with each other, as we had done with the human genome. I was confident that the genetic code was sufficiently unique for each species to enable its reconstitution in the computer out of a complex mixture of nonrelated sequences.

As it turned out, my new method of environmental shotgun sequencing was highly successful. We started with water samples from the Sargasso Sea surrounding Bermuda, which at that time was considered by many to be an ocean "desert," due to lack of available nutrients. In this "barren" sea we found in just one sample over 1.2 million previously unknown genes and at least 1,800 species.[12] The nutrient-poor seawater was full of life because the organisms' energy is derived directly from sunlight. We discovered that in addition to photosynthesis, almost every microorganism in the upper parts of the ocean has a light-sensitive protein, a rhodopsin photoreceptor, similar to the ones in our eyes. From our work sampling every two hundred miles as part of the *Sorcerer II* expedition, which has covered more than sixty thousand miles at sea over the past six years, we have discovered over eighty million genes. What was once thought to be single species are now known to be collections of thousands of closely related organisms. From numerous studies we have estimated there are on the order of 10^{30} single-celled organisms and 10^{31} viruses in the oceans. That amounts to a billion trillion organisms for every human on the planet.

The second line of research funded by the DOE enabled us to restart down the road toward the first creation of synthetic life. Clyde Hutchison, from his early work with Robert Sinsheimer and Fred Sanger, had suggested the bacteriophage phi X 174 as a test project. There were several reasons why it was an attractive target. The bacteriophage has a small genome, and phi X 174 can't tolerate many genetic changes, so it would also be a good test of synthesis accuracy. Also, there was a great deal of information about the virus, thanks to the efforts of Kornberg, Sanger, and, of course, Sinsheimer, who had been inspired to study phages by Max Delbrück and had, understandably, chosen one of the smallest he could find.[13]

To test a simple synthesis approach, in 1997 we made our first attempt to synthesize the phi X 174 genome from a series of overlapping fifty-base-pair oligonucleotides, followed by a polymerase chain reaction to make copies of the genome. Initially it looked as if our experiment had worked. Clyde Hutchison's laboratory notebook recorded how on April 22 we had produced DNA molecules of the right size, corresponding to the entire phi X 174 genome. The key question: was the synthesis accurate enough to produce the virus? Our plan was to use the synthetic DNA to infect *E. coli*. If the DNA contained no lethal errors, the bacterium would produce the coded proteins that would self-assemble to produce more copies of the phi X 174 virus. Unfortunately, our synthetic genome had no effect at all. We knew that DNA synthesis was error-prone, but our hope was that with this process of selection by infection we might find the one-in-a-million DNA strand with the correct sequence. That hope was dashed when the implications of our failure sank in. Even for a small virus, the ability to synthesize DNA accurately was going to be a much more demanding task than we had imagined, let alone the much grander undertaking of making an entire bacterial genome to produce a living cell. We strategized as a team about how to proceed and weighed the broader question of whether our ultimate goal of synthetic life was even possible. However, the opportunity to sequence the human genome would postpone some of these critical considerations for a few years. But when we returned to the synthetic-genome challenge in the wake of our success with the human-genome sequencing, we were determined to succeed.

Our initial attempts years earlier to synthesize phi X 174 had clearly failed because of the errors that inevitably occur during oligonucleotide synthesis. If automated DNA synthesizers were able to produce pure, error-free oligos of the preprogrammed sequences, then the assembly of long double-stranded DNA molecules would be relatively straightforward. But, in reality, only about half of the synthesized molecules have the correct chain length; the rest are mostly truncated molecules. The usual reason that they are shorter than expected is that the synthesizers will have a base that does not incorporate; this is called the N-1 problem. These incorrect molecules either halt the assembly of the oligonucleotides or result in assembled DNA with errors in the genetic code.

We did a simple calculation to figure out why we had not succeeded. Because on average only one of every two molecules used to assemble the phage DNA was correct, it became remarkably—and depressingly—clear why our assembly had failed years earlier. The probability that a strand of our phi X 174 genome would completely assemble correctly by random selection from 130 unpurified oligonucleotides was half to the power of 130: $(1/2)^{130}$, or 10^{-39}, a vanishingly small figure. We estimated that, even by exploiting selection by infectivity, it was essential to reduce the proportion of incorrect oligonucleotides to fewer than 10 percent of the population, because only then could we ensure that we had a chance to create enough correct molecules.

We went back to basics. First, we reevaluated the genome that we wanted to construct to make absolutely certain that we were beginning with an accurate viral sequence. We based that sequence exactly on the historic 1978 paper by Fred Sanger and colleagues. We were fortunate that Clyde Hutchison had kept a sample from the virus that had originally been sequenced, so that we could re-sequence it with the latest methods to test the accuracy of Sanger's team's work. We found only three differences out of the 5,384 base pairs, and it was not clear if these were due to errors in the original sequence or to viral variation on regrowth of the sample. Whatever the case, Sanger's sequence had proven to be remarkably accurate, a testament to his team's effort.

Sequence accuracy has always been an issue in the field of genomics. A substantial portion of early DNA sequencing was far less than 99 percent accurate (one error per one hundred bases). Only a few labs met a "high" standard—the one set for the human genome—of one error per ten thousand bases. The standard for writing genetic code is orders of magnitude higher than current standards for reading DNA software. Because digitized DNA sequences are going to be the basis of genome design and synthesis, the sequence they are based on needs to be extraordinarily precise if they are to lead to life. (Later on we discovered that a single "spelling mistake"—the deletion of just one base—out of 1.1 million letters of genetic code meant the difference between life and death, when it came to creating the first synthetic cell).

We have known since the work of Sinsheimer that the phi X 174 genome needs to be circular in order to be infective.[14] To create a functional circular genome synthetically, we broke the problem down into several steps. We started with the DNA sequence in a computer file and then divided the genome into overlapping pieces that were small enough to make with a DNA synthesizer. To synthesize the phage, we designed 259 oligos, each forty-two bases in length, which covered its genome in an overlapping fashion. The top strand of the genome would be formed from 130 of the oligos, and another 129 would form the bottom strand. Because the phi X 174 genome consists of 5,384 base pairs, the design also had to take into account the regions of overlap between the forty-two base pieces—which we called forty-two mers (from Greek *meros*, or "part," in reference to each base)—as well as additional sequences we added to each end of the genome to duplicate a restriction site contained only once in the genome, where the restriction enzyme PstI could cut the DNA to create overlapping ends that would bind to each other, causing the DNA to form a circle.

Knowing that only half of the synthesized forty-two-mer DNA fragments would be the correct length, we reasoned that by simply purifying the oligonucleotides we could greatly improve the assembly accuracy. DNA-sequencing gels sort DNA molecules of different lengths and are able to distinguish molecules only one nucleotide in difference. In the process, called gel electrophoresis, negatively charged nucleic-acid molecules move through an agarose gel under the influence of an electric field. The truncated oligos are smaller and, as a result, move faster than the correct-size oligos. By simply slicing the gel with a razor blade we were able to isolate the correct-size band that we wanted to use to assemble the top and bottom strands of phi X 174.

Now we had the components for constructing the phage genome in the form of purified oligo strands. We then pooled the top and bottom oligos, which, due to the overlapping design, aligned themselves in the proper order, again like self-assembling Lego blocks. We then permanently linked the fragments together with the enzyme DNA ligase. Rather than use the same enzyme that Kornberg did, we chose a more robust ligase enzyme, from a high-temperature organism, so that it would be active for long periods. After allowing the pool of oligos to react for eighteen hours at fifty-five degrees centigrade, we went from the 42 bases of oligos to assemblies with average sizes of around seven hundred bases, with some fragments measuring two thousand to three thousand bases.

We generated the full-length genome sequence of phi X 174 from these longer pieces of DNA using a method called polymerase cycling assembly, or PCA, a variant on polymerase chain reaction (PCR), a commonly used DNA-amplification method. With PCR we can amplify minute quantities of DNA by heating the sample, so the DNA denatures or melts, which separates double-stranded DNA into two pieces of single-stranded DNA. Next, the thermostable Taq polymerase makes two new strands of DNA, using the original strands as templates. This duplicates the original DNA, such that each of the new molecules contains one old and one new strand of DNA. Then each of these strands can be used to create two new copies, and so on and so forth.

In the variant of this process known as PCA, we start with all of the larger pieces of DNA from our first stage of assembly (averaging seven hundred base pairs). Once again we melt the double-stranded DNA into single strands. Instead of copying the single strands with DNA polymerase, we allow the reaction to cool, so that the single strands re-anneal with any complementary strands, exploiting the way that complementary bases always pair. The reason that this works is often two DNA strands will overlap only at one end, like aligning only the first joint of your index fingers, creating a much longer molecule. We then fill in the missing bases at each end using the DNA polymerase, which turns the single-stranded DNA into double-stranded DNA. By repeated cycling, you can build a piece of DNA several thousand base pairs long and do so relatively rapidly. The cycles continue until, from the random associations, the molecules grow to cover the entire genome. Finally, a regular polymerase chain reaction is used to amplify the entire genome sequence. To rejoin the ends of these linear phage genomes into an infectious circle, the amplified molecules were cut with the PstI enzyme, to leave end sequences that stick to each other, forming the circle.

Now came the important test to see if we had succeeded in creating an accurate, infectious synthetic genome. For an infectious virus to be produced, the synthetic DNA needs to be recognized by the enzyme systems in the *E. coli* cells, first by being transcribed into mRNA and then into the viral proteins by the *E. coli* protein-synthesis machinery. To make sure that our synthetic DNA was able to get into the target *E. coli* host bacterium, we used a method called electroporation, in which an electric field punches tiny temporary holes in the *E. coli* cell wall. After it was infected with the synthetic phi X 174, the *E. coli* was spread on agar, a jelly-like mixture of agarose and agaropectin, in a Petri dish and allowed to incubate at thirty-seven degrees centigrade for between six and eighteen hours.

We would be able to determine if the new strategy had worked if there was a plaque—a telling clear circle—on the lawn of *E. coli*. That would indicate that the viral proteins had been successfully produced in the bacterium and self-assembled to form sufficient copies of the phi X 174 virus, which would cause the host cells to burst open, release the virus, and infect the surrounding *E. coli* cells in the process. After he had opened the incubator, Ham phoned to ask me to come to the laboratory as soon as I could. When he showed me the first plate I was very pleased: there were clear plaques all over the plate. The synthetic bacteriophage DNA had indeed had the ability to infect, reproduce, and then kill the bacterial cells. Ham and Clyde were giddy with excitement. Our entire process to create the synthetic genome and infect the cells had taken us just two weeks.

To put our efforts in context, a more heroic effort to create a partially viable virus using a step-by-step process had been reported the year before by Eckard Wimmer, of the State University of New York, Stony Brook. His team had taken three years to produce the first

synthetic RNA virus by assembling the seven-thousand-base poliovirus genome from small synthetic DNA oligos guided by the sequence
that he and his colleagues announced in 1981. The synthetic DNA was
converted into infectious viral RNA using the enzyme RNA transcriptase. This first synthetic RNA polio virus suffered from the same inac-
curacy of oligo synthesis that had dogged our own efforts and, as a
result, had greatly attenuated activity.[15] The only negative aspect of
Wimmer's accomplishment was that he chose to publish it more as a
warning to the scientific community rather than a straightforward scientific result, raising the polemics and public concern.

In the wake of Wimmer's work, we had reduced the time required to
create a virus from years to days. Because this work had been funded
by the DOE, I contacted the department to notify the government of
our success. The official response was rapid, as I recounted in my autobiography, *A Life Decoded*:

> The very next day I found myself sitting in a restaurant a few blocks
> from the Oval Office on Pennsylvania Avenue where I had been sum-
> moned only two hours earlier to an urgent lunchtime meeting by Ari
> Patrinos, who worked at the Department of Energy's biology director-
> ate, which had sponsored the research, and who had also played an
> important role in the joint announcement of the first human genome
> at the White House. Among those who joined us was his boss, Ray-
> mond Lee Orbach, director of the Office of Science at the DOE; John
> H. Marburger III, the science advisor to the president and director of
> the Office of Science and Technology Policy; and Lawrence Kerr, di-
> rector of Bioterrorism, Research, and Development for the Office of
> Homeland Security at the White House. In the wake of the mailing of
> anthrax spores to public figures in October 2001, killing five people,
> the US Government had invested heavily to prepare for future acts of
> bioterror.[16]

I explained to them how quickly we had created phi X 174 with our
error-correction method and that now we could easily do so even

faster. Kerr looked apprehensive, and the questions raised by being able to create synthetic viruses eventually went all the way up to the White House, for a decision on potential restrictions on the publication of our results.

I had urged the government to review the same issue over a decade earlier, when my team, then at NIH, had been asked by the secretary of the Department of Health to sequence the smallpox genome as part of an international treaty set to destroy the remaining stores of the small-pox virus at the Centers for Disease Control, in Atlanta, and at a secure facility in Moscow. Whether or not to extinguish the remaining small-pox strains has been one of the fiercest debates in global public health in recent decades. The hope was that, if the genome could be sequenced prior to the destruction of the virus, then important scientific information would be preserved. The sequencing, as described in detail in *A Life Decoded*, started in my NIH lab and was finished at TIGR. As we analyzed the genome, we became concerned about several matters.

The first was whether the government would or should allow us to publish our sequencing and analysis. Our unease about releasing this knowledge was understandable: this virus has killed millions upon millions of humans. Before the HIV epidemic, the smallpox variola virus had been responsible for the loss of more human life throughout history than all other infectious agents combined. Thought to have originated over 3,000 years ago in India or Egypt, smallpox appeared in repeated epidemics that swept across continents, killing as many as 30 percent of those infected, or leaving the survivors disfigured and blinded. Smallpox is also believed to have wiped out a significant portion of the native populations of the Americas, where infected blankets were deliberately given by European settlers to indigenous populations to spread the disease.[17] Having also claimed the lives of kings, queens, tsars, and emperors, smallpox has altered the course of history.[18]

I eventually found myself in the National Institutes of Health, in Bethesda, with the director, Bernadine Healy (who died of a brain tumor in 2011), together with government officials from various agencies, including the department of defense. The group was very understandably worried about the open publication of the smallpox genome data. Some of the more extreme proposals included classifying my research and creating a security fence around my new institute building. It is unfortunate that the discussion did not progress to develop a well-thought-out long-term strategy. Instead the policy that was adopted was determined by the politics of the Cold War. As part of a treaty with the Soviet Union, which had been dissolved at the end of 1990, a minor strain of smallpox was being sequenced in Russia, while we were sequencing a major strain. Upon learning that the Russians were preparing to publish their genome data, I was urged by the government to rush our study to completion so that it would be published first, ending any intelligent discussion.

Unlike the earlier, expedient, thinking about smallpox, there was a very deliberate review of the implications of our synthetic-virus work by the Bush White House. After extensive consultations and research I was pleased that they came down on the side of open publication of our synthetic phi X 174 genome and associated methodology. We were fortunate that some of the funds for this first stage of our research had come from the government, since it ensured that there would be a rapid response from regulators. I knew that without public discussion and government review we could end up with a knee-jerk policy response driven more by the climate of fear that prevailed in the wake of the 9/11 attacks and Wimmer's polio virus, rather than by calm, clear-eyed logic and reason. The study would eventually appear in *Proceedings of the National Academy of Sciences* on December 23, 2003. One condition of publication from the government that I approved of was the creation of a committee with representatives from across government to be called the National Science Advisory Board for Biosecurity, (NSABB), which would focus on biotechnologies that had dual uses.

At the Washington, D.C., press conference that was arranged at the Department of Energy's headquarters to discuss the paper, the secretary of energy, Spencer Abraham, called the work "nothing short of amazing" and predicted that it could lead to the creation of designer microbes tailored to deal with pollution or to absorb excess carbon dioxide or even to meet future fuel needs. That, for me and society, would be the real prize. We now had the ability to construct synthetic genomes that I hoped would lead to extraordinary advances in our ability to engineer microorganisms for many vital energy and environmental purposes. Some, for example, could be used to turn sunlight into fuel, others to devour pollutants or scrub exhaust gases of carbon dioxide.

We had repeated, though this time using synthetic DNA, what Kornberg had achieved in the 1960s with a DNA polymerase copy of the then-unknown phi X 174 genome. These feats confirmed that the DNA code contained the necessary and sufficient information to make the virus: proof by synthesis. Having accurately made DNA fragments five thousand bases in size, we realized that we had solved a key limitation in DNA synthesis and could take the next step. We were now ready to attempt to go where no one had gone before, to create a whole bacterial synthetic genome and try to produce the first synthetic cell. Little did we realize that this would take us another seven years to accomplish.

Even then, however, we recognized that if we were successful in the ability to design the code of life in the computer, translate it into DNA software by chemical synthesis, and put that synthetic code to work to create a new organism, this meant that vitalism was truly dead and, as a corollary, that we would have a clearer picture of what the word "life" really meant. The fusion of the digital world of the machine and that of biology would open up remarkable possibilities for creating novel species and guiding future evolution. We had reached the remarkable point of being at the beginning of "effecting all things possible," and could genuinely achieve what Francis Bacon described as establishing dominion over nature. With this great power, however, came the duty to explain our purpose—so that society at large could understand it—and, above all else, to use such power responsibly.

Long before we finally succeeded in creating a synthetic genome, I was keen to carry out a full ethical review of what this accomplishment could mean for science and society. I was certain that some would view the creation of synthetic life as threatening, even frightening. They would wonder about the implications for humanity, health, and the environment. As part of the educational efforts of my institute I organized a distinguished seminar series at the National Academy of Sciences, in Washington, D.C., that featured a great diversity of well-known speakers, from Jared Diamond to Sydney Brenner. Because of my interest in bioethical issues, I also invited Arthur Caplan, then at the Center for Bioethics at the University of Pennsylvania, a very influential figure in health care and ethics, to deliver one of the lectures.

As with the other speakers, I took Art Caplan out to dinner after his lecture. During the meal I said something to the effect that, given the wide range of contemporary biomedical issues, he must have heard it all by this stage of his career. He responded that, yes, basically he had indeed. Had he dealt with the subject of creating new synthetic life forms in the laboratory? He looked surprised and admitted that it had definitely not been a topic he had heard of until I had raised the question. If I gave his group the necessary funding, would he be interested in carrying out such a review? Art was excited about taking on the topic of synthetic life. We subsequently agreed that my institute would fund his department to conduct a completely independent review of the implications of our efforts to create a synthetic cell.

Caplan and his team held a series of working groups and interviews, inviting input from a range of experts, religious leaders, and laypersons. I was invited to one session to describe the scientific approach that we had planned and to answer questions. I found myself sitting alongside representatives from several major religions. I was quite astonished, and pleased, when the discussion seemed to suggest that, because they could not find references in the Bible or other religious writings that forbade the creation of new life forms, it must be acceptable.

I did not hear again about the University of Pennsylvania bioethics study until the results were published in *Science*, in a paper entitled "Ethical Considerations in Synthesizing a Minimal Genome," coauthored by Mildred K. Cho, David Magnus, Arthur Caplan, Daniel

McGee, and the Ethics of Genomics Group.[19] (It was the same December 10, 1999, issue of *Science* in which our own study, "Global Transposon Mutagenesis and a Minimal Mycoplasma Genome," appeared, describing how we had used transposons to determine which genes were critical for life.) The authors hailed our work as an important step forward in genetic engineering that "would permit the creation of organisms (new and existing) simply from knowing the sequence of their genomes."

The paper began by pointing out how established thinking about ethics and law had been outpaced by the advances in science in the face of the surprise announcement of the cloning of the sheep Dolly in February 1997. (Dolly was not actually the first animal to be cloned but the first to be cloned from an adult cell.) This news came as a shock to biologists, because few thought it possible to take a differentiated adult cell and turn back the developmental clock to create an embryonic cell that could grow again into an animal. The sheep that donated the mammary cell to create Dolly was not "raised from the dead," as some have claimed.[20] Only her DNA software lived on.

As I had hoped, the Pennsylvania team seized the initiative when it came to examining the issues raised by the creation of a minimal genome. This was particularly important, in my view, because in this case it was the scientists involved in the basic research and in conceiving the ideas underlying these advances who had brought the issues forward—not angry or alarmed members of the public, protesting that they had not been consulted (although some marginal groups would later make that claim). The authors pointed out that, while the temptation to demonize our work might be irresistible, "the scientific community and the public can begin to understand what is at stake if efforts are made now to identify the nature of the science involved and to pinpoint key ethical, religious, and metaphysical questions so that debate can proceed apace with the science. The only reason for ethics to lag behind this line of research is if we choose to allow it to do so."[21]

The article went on to address a wide range of issues, including potential environmental dangers from the release of new species and considerations of patent law. One key paragraph on security concerns was largely overlooked in media reports, however, presumably because it was thought that the reality of synthesizing these genomes lay too far in the future: "The dangers of knowing the sequences of extremely deadly pathogens could pose threats to public health and safety that might outweigh the benefits. It is disturbing that current regulatory methods provide little if any oversight of these technologies."

Mindful of the debate about extinguishing smallpox and the concerns over the poliovirus publication, and perhaps anticipating a great deal of hand-wringing in future years about reviving pandemic organisms in influenza research,[22] the team had asked whether we should regulate the science and, if so, to what extent. Such questions would come to dominate the next iterations of synthetic-genome science.

Oddly, given that it had appeared in a scientific journal, the *Science* paper devoted a great deal of space to pondering the impact of reductionist science on "the meaning and origin of life" without really wrestling with the tricky issue of what was meant by that little four-letter word "life." The authors warned,

There is a serious danger that the identification and synthesis of minimal genomes will be presented by scientists, depicted in the press, or perceived by the public as proving that life is reducible to or nothing more than DNA.... This may threaten the view that life is special. At least since Aristotle, there has been a tradition that sees life as something more than merely physical. This provides the basis for belief in the interconnectedness of all living things and the sense that living things are, in some important way, more than organized matter.

As if to underline their anxiety on this matter, Cho et al. also focused a great deal of attention on religious issues: "Surprisingly, there has been little inclination within major Western religious communities to devise a definition of life or to describe the essence of life." Science, therefore, was left with that responsibility, even though the authors concluded that a "purely scientific definition of life" was cause for concern.

Perhaps the most pressing question, according to the Pennsylvania team, was "whether such research constitutes an unwarranted intrusion into matters best left to nature." An important—and, for me, reassuring—conclusion of the study echoed what I had heard in some of the earlier consultations, that "the dominant [religious] view is that while there are reasons for caution, there is nothing in the research agenda for creating a minimal genome that is automatically prohibited by legitimate religious considerations."

That did not mean that religious considerations were irrelevant, however. At one extreme was the view that our work marked progress for humankind. At the other was the view that it was only the most recent example of scientific hubris, one that would inevitably lead to catastrophe—a theme that had been examined and explored again and again in popular literature, from the monster of Mary Shelley's *Frankenstein* to the Beast People of H. G. Wells's *The Island of Doctor Moreau* to the revived dinosaurs of Michael Crichton's *Jurassic Park*.

It was this very issue that eleven years later would dominate the response of the press to our announcement of the first synthetic cell, when a chorus of questions followed one track of thought: weren't we "playing God?" Wisely, the Pennsylvania report pointed out that this objection had become a way of forestalling rather than fostering discussion about the morally responsible manipulation of life. It argued that it was possible to strike a balance between the pessimistic view of these efforts as yet another example of hubris and the optimistic view of their being tantamount to "human progress." The authors added that a "good steward" would move genomic research forward with caution, armed with insights from value traditions with respect to the proper purposes and uses of new knowledge. They concluded that there were no strong ethical reasons that should prevent the team from continuing research in this field as long as they continue to engage in public discussions; which we do.

6 First Synthetic Genome

The present machines are to the future as the early Saurians to man.

—Samuel Butler, 1872[1]

The quest to manipulate life in the laboratory has come a long way since the early days of recombinant DNA, in the 1970s, when Paul Berg, Herbert Boyer, and Stanley Cohen began to cut and splice DNA. By the end of that decade a laboratory strain of *E. coli* had been genetically altered to produce human insulin. Since then scientists have induced bacteria to manufacture human clotting factors to treat hemophilia and to make growth hormone to treat dwarfism. In agriculture, DNA has been altered to make plants resistant to drought, pests, herbicides, and viruses; to boost their yields and nutritional value; to manufacture plastics;[2] and to reduce the use of fossil-fuel-based fertilizer. Animal genes have been altered in attempts to increase yields, to produce models of human disease, to make drugs such as anticoagulants, and to produce "humanized" milk, and pig organs that can be transplanted into people. Genetically modified cells have been used to manufacture proteins, from antibodies to erythropoietin that increases the production of red blood cells. Some patients have been genetically altered via gene therapy, in which a software "patch" is used to treat genetic disorders such as immune deficiency, blindness, and the inherited blood condition beta-thalassemia.

Genetic engineering has today evolved to be more commonly known as synthetic biology. The distinction between molecular biology and synthetic biology is blurred, and in most uses there is no actual distinction. "Synthetic biology" just sounds sexier, and in the same way, "systems biology" has replaced physiology, and some good old-fashioned chemists like to rebrand their efforts as nanotechnology. Whatever you want to call it, around the globe large numbers of scientists are pursuing genetic engineering by blending biology with engineering approaches.

Recent achievements are too numerous to list in detail, but here are a few examples of genetic-engineering advances. The workhorse of molecular-biology labs, *E. coli*, was partly minimized (with 15 percent of its DNA removed) by Frederick Blattner, at the University of Wisconsin, in 2002[3] in an attempt to make it a more reliable chassis for industrial production. At Harvard, George Church's lab developed a method called MAGE—multiplex automated genome engineering—to replace a codon in thirty-two strains of *E. coli*, then planned to coax those partially edited strains along an evolutionary path toward a single-cell line in which most instances of the codon had been replaced.[4] In Christopher Voigt's laboratory, at MIT, a sophisticated genetic circuit has been assembled that, if installed in a bacterium, could, for example, enable it to sense four different cancer indicators and, in the presence of all four, release a tumor-killing factor.[5] His colleague, Timothy Lu, has developed DNA modules that can perform logic operations, so that programmable cells with decision-making capabilities could be customized for multiple applications.[6] Inevitably, as technology advances, the goals become more ambitious, allowing new questions to be approached that in turn result in further technological development.

Our proposed effort to synthesize the genome of a living cell in order to understand the genes required for life would require substantial advances. To rise to the challenge we needed to draw on a diverse range of skills, as we had successfully done earlier with the sequencing of the human genome; success in modern science is increasingly dependent upon good teamwork.[7] In order to create a synthetic cell we initiated three major programs. Based on our work with phi X 174 we all assumed that DNA synthesis was the area that required the most focus if we were going to be successful. Therefore the first team, the synthetic-DNA team, had the task to synthesize the entire bacterial chromosome. This group was headed by Ham Smith and included

Daniel G. Gibson, Gwynedd A. Benders, Cynthia Andrews-Pfannkoch, Evgeniya A. Denisova, Holly Baden-Tillson, Jayshree Zaveri, Timothy B. Stockwell, Anushka Brownley, David W. Thomas, Mikkel A. Algire, Chuck Merryman, Lei Young, Vladimir N. Noskov, John I. Glass, and Clyde A. Hutchison III. I was certain that the chemistry was solvable, and I was much more concerned about the biology. Could we transplant and boot up a synthetic genome if we succeeded in synthesizing one, and could we do more to understand which genes might be essential for minimal life? The second and third teams were therefore focused on biology. The genome-transplant team was headed by John I. Glass and included Carole Lartigue, Nina Alperovich, Rembert Pieper, and Prashanth P. Parmar. The minimal-gene team was led by John I. Glass and Clyde Hutchison, and included Nacyra Assad-Garcia, Nina Alperovich, Shibu Yooseph, Matthew R. Lewis, and Mahir Maruf. While there were overlaps in all three teams, each was intensely focused. Ham, Clyde, and I were the overall project leaders, and when the Venter Institute, in La Jolla, was established and Ham and Clyde moved west, John Glass played a broad leadership role in Rockville.

Our plan was to synthesize the smallest known genome that can constitute a living self-replicating cell, *M. genitalium*. The thinking was that the DNA synthesis would be the greatest challenge and that it would provide us with the means to reduce the small genome even further, in order to understand and dissect the genetic instruction set of a simple cell, so that we might see and understand the minimal gene set for life. For the genome synthesis we divided the *M. genitalium* genome into one hundred and one segments that we called "cassettes," each around the size of the phi X 174 genome. We knew that we could accurately make synthetic DNA in five thousand to seven thousand base-pair segments, and our plan was to find a way to combine them to reconstruct the *M. genitalium* genome. The 582,970-base-pair *M. genitalium* genome was twenty times larger than anything that had been synthesized before. Prior to this effort, the largest synthetic DNA constructs included two small viruses and the 32,000-base-pair "polyketide gene cluster." (Polyketides are ring-like chemicals made naturally by bacteria, fungi, plants, and marine animals to kill predators, and have been used as the basis of many drugs, notably antibiotics and anticancer agents.)[8]

We needed, therefore, to develop a new set of tools for the reliable synthesis of large DNA molecules. Tool and technique development is at the heart of scientific advancement, but high scientific standards are, in my view, just as essential. I have often described laboratory work in genomics as a "garbage in and garbage out" process, just like computing, in that if extreme diligence is not exercised every step of the way, the end result will, at best, be less than a quality outcome. When we sequenced the first genomes in the 1990s we found that if our DNA libraries (containing small fragments of the genome) were not of the highest quality, and did not represent a true random distribution of all the DNA in a genome of interest, then it was highly unlikely that a computer could use the sequences generated from those library samples to reconstruct the genome sequence. The same applied to the quality of the DNA used for sequencing, the purity of the reagents, and the reproducibility of the techniques. All had to be of the highest standard. My teams paid particular attention to these basics, and as a result we were consistently able to generate very high-quality DNA-sequence data.

However, as discussed in Chapter 5, the quality of DNA-sequence data necessary to read genetic code is much lower than that required to write a life-supporting code. The target set by the genome community for the former was less than one error per ten thousand base pairs. While that might sound like a low error rate, using that standard would mean that we would have about sixty errors in the *M. genitalium* genome and more than sixty thousand errors in the human genome. Clearly, such compromised data would be unlikely to support life and clearly could not provide sequence information of sufficient quality for the accurate diagnosis of human genetic changes associated with disease. A typical human gene could be spread over thousands to millions of base pairs, so current error rates would mean multiple sequence errors per gene. To put this in context, a single error in a gene can cause a serious disease, such as the blood disorder sickle-cell anemia. By the same token, this error rate was unlikely to be low enough for the reconstruction of a genome to create a living cell.

These simple facts are often overlooked in the fanciful discussions about reviving extinct species from their sequenced genomes. Whether inspired by the great advances in paleogenomics, Svante Pääbo's sequencing of the Neanderthal genome,[9] or the sequencing of woolly mammoth DNA at Pennsylvania State University,[10] journalistic speculation always turns to the fevered contemplation of species resurrection.[11] I have read too many articles that breezily discuss the reconstruction of a Neanderthal or a woolly mammoth with the help of cloning, even though the DNA sequences that have been obtained for each are highly fragmented, do not cover the entire genome, and—as a result of being so degraded—are substantially less accurate than what is routinely obtained from fresh DNA.

Reading the Neanderthal DNA has, however, been a wonderful scientific advance that has taught us much about our own evolution, revealing that the interbreeding of some ancestors of modern humans with our Neanderthal cousins has left us with a legacy of 3 to 4 percent of their genome originating with the Neanderthals.

In order to synthesize the *M. genitalium* genome, we required an extremely accurate DNA sequence. Our effort to sequence the first two genomes, in 1995, relied on early versions of DNA sequencers, and while we had accuracy better than the one error in ten thousand, we were concerned that the sequence was probably not of sufficient quality to produce data of sufficient precision to generate a living cell. We had no option other than to re-sequence the *M. genitalium* genome using the latest technology. Our new sequence revealed that our original version was accurate to the level of one error per thirty thousand base pairs, and when the old and new sequences were combined, we had less than one error per hundred thousand base pairs—around half a dozen errors in the entire genome. It was from this new, highly accurate sequence that we began to design the synthesis of the *M. genitalium* genome.

Our success with converting the digital code of phi X 174 into chemical DNA gave us the confidence to tackle the much larger genome of a free-living organism. Since we could produce viral genome-size segments with high accuracy, we knew we had a chance of success if we could break down a bacterial chromosome into segments of that size and find a reliable way to stitch them all together.

We carved up the target genome into 101 cassettes that ranged in size from five thousand to seven thousand base pairs. The cassettes were designed to overlap with neighboring cassettes by at least eighty base pairs, with the longest overlap being 360 base pairs, so that we could link them like Legos. We designed our cassettes so that the DNA sequences were complementary in the overlap zones: if the final letter on one cassette was a T, it would want to bind to an A on the other. Rather like a zipper, the overlaps knit together the complementary bases to form the helix.

In our effort to create the synthetic genome, there were two more considerations to take into account. The *M. genitalium* genome, like the phi X 174 genome, is circular, so we designed cassette 101 to overlap with cassette number one. As part of our genome design, we also wanted to have a foolproof way of distinguishing our synthetic genome from the native *M. genitalium* genome. This was essential to prevent artifacts from misleading us, so that we could always track the synthetic genome and prove without a doubt that it was driving a new synthetic cell, not a contaminating native cell or genome.

Rather like artists sign their work, we wanted to leave a signature in the new genome to distinguish it from a natural genome. So, using single amino-acid code abbreviations, we designed "watermark" sequences that would spell out "Venter Institute" and "Synthetic Genomics," as well as the names of the key scientists working on the project. We used a different codon to represent each of twenty letters in the alphabet (not all letters were represented, so *v* was used interchangeably with *u*, for example). My name, when coded this way, came out

TTAACTAGCTAATGTCGTGCAATTGGAGT
AGAGAACACAGAACGATTAACTAGCTAA

These watermark sequences were inserted into five different cassettes spaced throughout the genome. We also needed to insert an antibiotic resistance gene that would enable us to selectively kill cells lacking our new genome and thus enable us to select for our synthetic genome. As part of the genome design, we put the antibiotic-resistance gene into a key *M. genitalium* gene, MG408, which is required for the bacteria to adhere to mammalian cells. Because this gene plays a role in the ability of the organism to cause disease, we effectively crippled it and ensured that the synthetic organism would be harmless.

So that our team could concentrate on the key step, which was assembling the 101 cassettes into a genome, I insisted that we invite three DNA-synthesis companies to bid for the contract to make the 101 designed cassettes for us. Despite the advertising claims, we found only one company that could make five thousand to seven thousand base-pair segments. Also, this was an expensive proposition: DNA synthesis costs on the order of $1 for each base pair, so our raw material alone was going to cost more than half a million dollars. Having made such a serious financial commitment, we were determined to make it work.

One of our biggest challenges would be how to link the 101 cassettes. One idea came out of our early genome-sequencing projects. As a result of our attempts to cover as much biological diversity as possible, I had become aware of a quite remarkable organism that was able to rebuild its genome after suffering substantial radiation damage. In 1999 we published the paper "Complete Genome Sequencing of the Radio-resistant Bacterium, *Deinococcus radiodurans* R1,"[12] which described at a genome level a novel organism that could withstand up to three million rads of ionizing radiation. Given that exposure to only five hundred rads of the same radiation is lethal to humans, how did *Deinococcus* survive such an assault? Could we harness the same DNA-repair mechanisms to build a synthetic genome?

Radiation has the same effect on the proteins and DNA of all species, which is related in part to the size of the molecules. Early in my science career, I devoted considerable time to using radiation to inactivate proteins in order to determine their size. The technique is easy in principal. In proteins, radiation breaks the peptide bonds that join the constituent amino acids together; one hit per protein is sufficient to destroy its activity. There is an inverse relationship between the size of a protein molecule and the dose of radiation needed to inactivate it by breaking peptide bonds (larger targets have a much greater chance of being hit than small ones) therefore the smaller the protein, the greater the dose of radiation that is required. I used the method to determine the size of neurotransmitter receptor proteins and their functional complexes.[13]

Radiation affects DNA in a similar way, breaking the chemical bonds that link the bases together. As with proteins, the larger the genome, the lower the dose of radiation required to cause damage. Due to our larger genome, humans are much more sensitive to the effects of radiation than bacteria are. The genome of a human cell, compared with that of a microorganism, is more than a thousand times greater in size: six billion base pairs, versus one to eight million base pairs for bacteria. As a consequence, it would take a much smaller dose of radiation to cause a double-stranded break in our DNA than it would for that of a bacterial chromosome. For that reason, we can be confident that, if we were unfortunate enough to suffer nuclear Armageddon, smaller life forms would endure.

So how does *Deinococcus* survive? When exposed to millions of rads of radiation, the *Deinococcus* genome shatters, with hundreds of double-stranded DNA breaks, but it can repair and reassemble its chromosomes and continue replicating. Its ability to do so is still not completely understood but involves, in part, having multiple copies of each chromosome, so that when it does experience random radiation-induced breaks in its DNA, the fragments can self-align to produce a template for DNA repair. I have often likened this process to the one that we use in shotgun sequencing, when software running on powerful computers reassembles randomly sequenced overlapping DNA fragments to reconstruct a genome.

We reasoned that if we could reproduce the *Deinococcus* DNA repair and chromosome-assembly process outside of *Deinococcus* cells, then we might be able to use it to assemble our synthetic chromosome from large, viral-size DNA segments. Two of our staff scientists, Sanjay Vashee and Ray-Yuan Chuang, agreed to take on this job. The team sorted through the *Deinococcus* genome for all the genes that might be relevant, then spent another two years cloning each gene, so that they could produce the repair proteins in the laboratory, where they could try a range of combinations in order to recapitulate the DNA assembly and repair. After a tremendous effort, we were forced to give up. We had hit a dead end and needed a new strategy.

Our next approach was to develop a logical, stepwise assembly plan. Making use of the designed overlaps in the DNA sequence of adjacent cassettes, we would assemble two cassettes in vitro to form a bigger fragment. Then we would clone the new larger fragment in *E. coli* so that, as this organism multiplied, so did copies of the larger fragment. In this way, we would produce sufficient DNA for the next stage in the assembly. Our goal, ultimately, was not only to produce the *M. genitalium* genome but also to establish a robust, reproducible assembly process that we could apply to create all sorts of synthetic genomes in years to come.

For the first round in the assembly of the genome, our plan was to link four cassettes, each of which was around the size of the phi X 174 genome, to create twenty-four thousand base-pair assemblies. This was accomplished by adding equal amounts of each of the four cassettes to a microfuge (micro-centrifuge) tube with a DNA vector that would enable us to multiply this newly constructed genome segment in *E. coli*. The DNA vector that we used is called a bacterial artificial chromosome (BAC), where one end of the BAC overlapped with the start of cassette number one, and the other end with the end of cassette number four.

To knit the pieces together, we added to the DNA mixture in the tube an enzyme (known as a 3'-exonuclease) that chews the end of the DNA, digesting only one of the DNA's two strands (called the 3' strand, a reference to how the carbon atoms are numbered in the sugars in the DNA's nucleotides) and leaving the other strand (the 5' strand) exposed. Using changes in temperature to control the exonuclease, we could ensure that the corresponding single-stranded ends of the cassettes would find each other and stick together, due to the chemical attraction of the complementary bases on each strand, à la Watson and Crick.

To ensure that we ended up with complete double-helix strands, we then added DNA polymerase, as well as some free nucleotides, so that any place where the 3'-exonuclease chewed away too much of the strand, the polymerase would fill in the missing bases. Next, another enzyme, DNA ligase, was added to the mixture to link the overlapping strands together. When all the enzymes had finished their work, we ended up with all four cassettes linked to form the 24,000-base-pair, or "24kb," strand. To produce all of the 24kb cassettes that together comprised the entire *M. genitalium* genome, we repeated this process twenty-five times.

Because we had reproduced the synthetic DNA in *E. coli*, we had sufficient DNA for sequencing. Following sequence verification of all twenty-five cassettes we repeated the in vitro process, this time linking three of the 24kb cassettes to form 72,000-base-pair cassettes, each equivalent to around one eighth of the *M. genitalium* genome. To do this, we first had to liberate (using a restriction enzyme) the 24kb cassettes from the BAC vector that was used to grow them in the *E. coli*.

Our BAC vectors were designed so that they contained an eight-base sequence on both sides of our inserted synthetic DNA. This eight-base sequence, which does not occur naturally in the *M. genitalium* genome, is recognized by a particular restriction enzyme, called NotI. When NotI cuts the BAC DNA, the 24kb synthetic fragment is released. At this stage we had achieved synthetic DNA more than twice the size of the previous record set for a synthetic-DNA assembly.

The next step was to repeat the process once again, this time to produce segments of 144,000 base pairs, each equivalent to one quarter of the genome. This was accomplished by subjecting two of the 72kb cassettes to the same in vitro assembly process. At this point, however, we were entering uncharted territory and pushing our techniques to the limit. As we reached the penultimate step—producing half-genome segments of 290,000 base pairs by combining two of the four quarter-segments together—we ran into problems: the 290kb segments appeared too large for *E. coli* to accommodate.

The team therefore began a search for other species that might be able to stably accommodate these large synthetic DNA molecules. We looked at *B. subtilis*, which had been used by a Japanese team to grow large segments of a bacterial algae genome.[14] But while *B. subtilis* could indeed accommodate the large 290kb segments, there was no way to recover the DNA intact from these cells, so we looked elsewhere. The solution came from the more complex cellular world of the eukaryote and a favorite experimental subject of scientists around the world studying eukaryotic biology: brewer's yeast, *Saccharomyces cerevisiae*. For centuries *S. cerevisiae* has been used for alcohol fermentation as well as for making bread, but in the laboratory it has been routinely exploited because it has a relatively small genome and an array of genetic tools that make genetic manipulation easy. For example, *S. cerevisiae* uses what is called homologous recombination, in which segments of DNA with sequences on its ends similar or identical to those in the *S. cerevisiae* genome can be spliced into its genome, replacing the intervening sequence.

Yeast cells, which are around ten times larger than those of *E. coli*, are protected by a thick cell wall, which is a barrier to DNA transformation of the cells. To overcome this, yeast cloning relies on the use of an enzyme, zymolyase, to digest much of the cell wall and create what are called spheroplasts, into which the large pieces of DNA can be more easily introduced.[15] Yeast cloning results in circular artificial chromosomes that appear stable and, due to their circular nature, have the advantage of being easily purified from the normal linear chromosomes of yeast.

We found that by using the yeast cloning, we could stably grow our large synthetic DNA constructs, and by using the yeast homologous-recombination system, we could link our overlapping one-quarter genome segments to form one-half genome pieces. This system then allowed us to assemble the entire *M. genitalium* genome in yeast. The end of the long, hard climb to the first synthetic genome of a living organism now seemed to be in sight.

We introduced six pieces of DNA into the yeast cells: the yeast-cloning vector and five that corresponded to the *M. genitalium* genome (the four one-quarter synthetic-genome segments, with one of the synthetic-DNA segments being split in two to overlap the yeast cloning sites). For this experiment to work, the yeast cell needed to take up all six DNA segments and put them together by homologous recombination. We screened ninety-four transformed yeast cells for DNA of the correct size and found that seventeen contained a complete synthetic *M. genitalium* genome.

Although it appeared that we had indeed succeeded in the assembly of our synthetic bacterial genome in yeast cells, we needed DNA sequencing to check the accuracy of the synthetic genome and ensure that the assembly process did not cause any errors. While that sounds simple, we had to develop new methods to recover our synthetic chromosome from the yeast cells, which we estimated represented only 5 percent of the total DNA in the cells. To enrich for our synthetic DNA, we used our knowledge of the genome sequence of the yeast and of the synthetic genomes to select for restriction enzymes that would only cut the yeast DNA into small pieces. We then used gel electrophoresis to separate the splintered remains of the yeast DNA from that of the intact synthetic chromosome.

Finally, we could use our whole genome shotgun method to sequence the synthetic genome. We were all very pleased and relieved when the DNA sequence exactly matched our computer-designed sequence, including the watermarks we had introduced. We had synthesized a 582,970-base-pair *M. genitalium* genome, and in achieving that feat we had created the largest synthesized chemical molecule with a defined structure.

We called our first synthetic chromosome *M. genitalium* JCVI-1.0. We wrote up our results and submitted them to *Science* on October 15, the day after my 61st birthday. Our paper was published online on January 24, 2008, and appeared in print on February 29. We celebrated our success in creating the genome but knew that our biggest challenges were yet to come: we now had to find a way to transplant the first synthetic genome into a cell, to see if it could function like a normal chromosome. In the process, the host cell would be transformed into one where all the components were manufactured according to the instructions held in our synthetic DNA. Once again, our efforts would build on earlier work and ideas that had originated from a range of talented teams, stretching back over many decades.

7 Converting One Species into Another

The transition from a paradigm in crisis to a new one from which a new tradition of normal science can emerge is far from a cumulative process, one achieved by an articulation or extension of the old paradigm. Rather it is a reconstruction of the field from new fundamentals, a reconstruction that changes some of the field's most elementary theoretical generalizations as well as many of its paradigm methods and applications. During the transition period there will be a large but never complete overlap between the problems that can be solved by the old and by the new paradigm. But there will also be a decisive difference in the modes of solution. When the transition is complete, the profession will have changed its view of the field, its methods, and its goals.

—Thomas Kuhn, 1962[1]

If I were to select a single study, paper, or experimental result that has influenced my understanding of life more than any other, I would without any doubt choose one above all others: "Genome Transplantation in Bacteria: Changing One Species to Another."[2] The research that led to this 2007 paper in *Science* not only shaped my view of life but also laid the groundwork that made it possible to create the first synthetic cell. Genome transplantation not only provided a way to carry out a striking transformation but would also help prove that DNA is the software of life.

There was some conceptual precedent for our efforts in a process called "nuclear transfer," which was used, for example, to create Dolly,[3] the cloned sheep, by a team led by Ian Wilmut, at the Roslin Institute, near Edinburgh, Scotland. The nucleus containing DNA from an adult sheep's mammary cell was transplanted into an egg that had been emptied of its own nucleus, effectively returning the mammary-cell DNA to an embryonic state. The resulting birth of Dolly made headlines worldwide in 1997 because she was created from an adult mammary cell. (Hence her name, a nod toward that well-endowed singer Dolly Parton.) Until her birth, it had been thought impossible to take a cell from an adult and produce a clone. The achievement of the Roslin Institute rested on many factors, from a technical understanding of the cell cycle to practical considerations, such as protecting reconstructing embryos in a shell of protective agar.[4] But Dolly was far from being the first clone and was not the first cloned sheep, either, as many believe.[5]

The history of nuclear transfer actually dates back to 1938 and the highly creative and influential German embryologist Hans Spemann (1869–1941), who published the first nuclear-transplantation experiments.[6] Spemann was the pioneer of what he called *Entwicklungsmechanik*, or "developmental mechanics," and was awarded the Nobel Prize for his efforts in 1935. With Hilde Mangold (1898–1924) he conducted the first nuclear-transfer experiments on the newt, which was an ideal experimental subject because of its large, easily manipulated eggs. In 1938 Spemann published the milestone text *Embryonic Development and Induction*, which described how his experiment rested on the dexterous use of microscopy, tweezers, and a delicate hair, probably plucked from his daughter Margrette.

Spemann used the hair as a noose to divide the cytoplasm of a newly fertilized salamander egg under the gaze of a binocular microscope, creating a dumbbell-shaped embryo. On one side of the dumbbell was the nucleus, containing DNA; on the other was only cytoplasm, which contains all of the contents outside of the nucleus that are enclosed within the cell membrane. (It is important to remember that, although Spemann had first been inspired by August Weismann's work on heredity, all that was known was that the secret of inheritance lay in the nucleus.) After the nucleated side divided four times, to grow into a sixteen-cell embryo, Spemann untied the hair and allowed one of the sixteen nuclei to pass into the separated cytoplasm in the other half of the dumbbell, forming a new cell with the original contents of the egg, along with a more mature nucleus. He once again tightened the hair and separated the two embryos. By doing this, he demonstrated that the nucleus still had the ability to turn into any type of cell after undergoing four divisions. In this way Spemann had created a clone, a genetically identical copy that was a few moments younger than the other.

Spemann called the process twinning, and it has been recorded in history as the first animal clone produced in the laboratory by nuclear transfer. He wanted to go further and had proposed a "fantastical experiment" to do the same with an adult cell, but like so many before him, he remained sufficiently overwhelmed and awed by the mysteries of development that he suspected it depended on more than mere physics and chemistry.

In the following decade Spemann's challenge attracted the attention of Robert Briggs, a researcher at Lankenau Hospital Research Institute, in Philadelphia (later the Institute for Cancer Research and now the Fox Chase Cancer Center), who was studying the cell nucleus. In 1952, working with Thomas J. King, he cloned leopard frogs using a method of nuclear transfer. Briggs and King's experiment was similar to those envisioned—and piloted using salamanders—by Spemann in his 1938 proposal. They transferred the nucleus of a frog cell from an early embryo into the large (one-millimeter) egg of the common American leopard frog, creating frog embryos that formed tadpoles. But they concluded in further experiments that developmental potential diminishes as cells differentiate, and that it was impossible to produce a clone from the nucleus of an adult cell. Then, working in Oxford in 1962, John Gurdon replaced the cell nucleus of a *Xenopus* frog's egg cell with one from a mature, specialized cell derived from the intestine of a tadpole. The egg developed into a cloned tadpole, and subsequent experiments yielded adult frogs.[7] His research taught us that the nucleus of a mature, specialized cell can be returned to an immature state and this led to his being awarded the Nobel Prize for Physiology or Medicine in 2012,[8] fifty years after his pioneering experiment.

What we were attempting to do was in some ways much more complex than these pioneering nuclear-transfer experiments, remarkable though they undoubtedly were. The work of Spemann was a little like working out how to reprogram a computer without knowing anything about software, by simply downloading code from the Web. Unlike more complex eukaryotic cells, bacteria lack a nucleus, a cellular substructure enclosed in a membrane. In the bacterial cell the genome is suspended in the cytoplasm's thick soup, along with the rest of the cell's components. As a result there was no cell organelle that could be surgically removed. An even greater potential challenge we faced was that we wanted to transplant the genetic material from one species to another, while all the previous nuclear-transfer experiments had involved working within the same species, if not the same exact animal.

When we began considering how to transfer synthetic DNA into a bacterium and replace its existing chromosome, we knew that we needed to develop a new method of genome transplantation, since the entire genome of the host species had to be replaced by inserted naked DNA of the genome of the new species without any blending (recombination) of the two genomes. For decades individual genes have been transferred as a matter of routine in molecular biology without limitations, including viral genes and human genes being transplanted into bacteria and yeast, where they are expressed. Yet no one to my knowledge had even attempted to transplant an entire genome, a task that many may have viewed as impossible.

Such biases often limit our ability to attempt new approaches or to accept new breakthroughs. For example, microbiologists once thought that bacterial cells could have only a single chromosome. But the reality was much more interesting, as I discovered in the mid-1990s, when we[9] sequenced the genome of cholera, a major disease that affects five million people worldwide, leading to as many as one hundred twenty thousand deaths every year.[10] One unique aspect of the algorithms that we had developed for shotgun sequencing, in which computers were used to match up overlapping pieces of sequenced code, was that they assembled genome sequences only on the basis of the identity of the overlapping genetic code. They had no preconceived notion of how many chromosomes, plasmids, or viruses were being assembled other than to put matching fragments together in a mathematically valid manner. When we assembled the sequenced fragments from the cholera genome, they cleanly assembled into two independent chromosomes—not one, as most had assumed. When we compared the two chromosomes with each other and with other genomes, we found that they were very different from each other.

After we made that discovery in cholera, we identified a significant number of microbial species with multiple chromosomes. That begged a question: how did these species acquire these multiple chromosomes? Did a cell just take up the additional DNA by chance from a lysed cell and the new chromosome became established because it added some key survival capabilities to its new home? Had two ancient bacterial cells fused to form a new species? We did not have the answers, but to me these ideas had great appeal. The evolution of species was thought by most to be due to the gradual accumulation of single base changes in the DNA sequence over millions to billions of years; the species in question adapted to its environment if the random changes offered a survival advantage. It made sense to me that some of the dramatic leaps we have seen in evolution were due at least in part to the acquiring of an additional chromosome that added thousands of genes and new traits in a single event.

We now know that some of the advanced functions of eukaryotic cells arose in evolution when early eukaryotic cells engulfed completely some microbial species that initially lived in a symbiotic relationship with them. Arguably the most important example occurred some two billion years ago, when a eukaryotic cell acquired a photosynthetic bacterial algae cell, which ultimately became in all plants the chloroplast, where photosynthesis occurs. The second most vivid example of this process, known as endosymbiosis, can be found in the "power packs" of our cells, mitochondria—which, like chloroplasts, carry their own genetic code and were derived from a symbiotic *Rickettsia* bacterium.

Because it was clear that transplanting genomes and entire cells was an essential part of our evolution, I was confident that we could find a way to artificially transplant genomes. For our initial transplantation experiments we chose mycoplasmas, because unlike most bacteria they lack a cell wall (a tough, fairly rigid outer layer) and have only a cell lipid membrane, which would simplify moving DNA into the cell. We also had extensive data sets including the *Mycoplasma* genome sequences and extensive gene-knockout data. I formed a new transplantation team around two scientists: John Glass, who had spent much of his career at Lilly pharmaceuticals working on mycoplasmas and had joined us when Lilly shut down its antimicrobial program; and a new French postdoctoral fellow, Carole Lartigue, who had had experience working with mycoplasma species. Other key members of the team included Nina Alperovich and Rembert Pieper.

We had first wanted to try to transplant the mycoplasma genome into *E. coli*, but while the mycoplasmas use the same four bases in their DNA as all other species, the codon UGA codes for tryptophan in mycoplasmas, whereas in other species UGA codes for a stop codon, which terminates the process of transcribing DNA into a protein. Because *E. coli* would read UGA as a stop codon, it would produce truncated proteins incompatible with making a viable cell. We would have to use a second mycoplasma species for the transplant experiments.

We had selected *M. genitalium* for genome sequencing and analysis because of its extremely small genome and, by the same token, for DNA synthesis, because we thought that the limiting step in making a synthetic genome would be our ability to use chemistry to reproduce the genome. That decision would be one we would come to rue, however, for an important practical reason: in the laboratory *M. genitalium* grows very slowly. While *E. coli* divides into daughter cells every twenty minutes, *M. genitalium* requires twelve hours to make a copy of itself. While that might not sound like a big difference, with logarithmic growth it is the difference between having an experimental result in twenty-four hours versus several weeks. Therefore, for the initial genome-transplant procedures we chose two different fast-growing mycoplasma species, *M. mycoides* and *M. capricolum,* which are both opportunistic pathogens of goats and probably took hold when the animals were domesticated, about ten thousand years ago.[11] Being significant livestock disease agents, these organisms are grown routinely in research labs. While not as fast as *E. coli, M. mycoides* divides in sixty minutes, and *M. capricolum* divides in one hundred minutes.

Being a genome lab we sequenced the genomes of both species in order to learn just how related they were. We found that *M. mycoides* has a 1,083,241-base-pair genome, of which three quarters (76.4 percent) matched in sequence to the slightly smaller *M. capricolum* genome, which consists of 1,010,023 base pairs. In the genome regions where the sequences align, they match with 91.5 percent accuracy. The remaining quarter (24 percent) of the *M. mycoides* genome that does not correspond in sequence to its smaller sister organism contains sequences that are not found in the *M. capricolum* genome.

We reasoned that, based on the DNA-sequence similarity, the key proteins in each species used to interpret the instructions in genetic code would have sufficiently compatible biochemistry to read the other's genome, while at the same time being sufficiently genetically distinct that we would be able to distinguish one from the other easily. We also thought that the genome sequences were different enough to be unlikely to recombine.

When it came to selecting which genome to be transplanted and which to be the host, we chose *M. mycoides* as the donor genome and *M. capricolum* as the recipient: *M. mycoides* grows faster and has the larger genome, which would make successful transplants easier to detect. There was also a technical reason that influenced our choice. In her previous lab, Carole Lartigue had studied a special stretch of DNA software known as the "origin of replication complex," or ORC, one that is central to the process of cell division. She had shown that, as a result of their respective ORCs, plasmids bearing the *M. mycoides* origin could grow in *M. capricolum*—but not vice versa.

Once we selected the donor and recipient mycoplasmas, we felt confident that we had a good experimental system that would offer the best opportunity to try genome transplantation. The next steps required a lot of tedious trial-and-error experiments and would take us into unknown territory. We had to develop new methods to isolate the intact chromosome from one species and transplant it into a recipient cell without damaging or cutting the DNA. While small pieces of DNA are easily manipulated without damage, large ones, especially entire chromosomes consisting of millions of bases, are very brittle and easily damaged.

We also had to decide on an overall experimental approach. Were we going to try to remove or destroy the *capricolum* genome before we transplanted in a new one from the *mycoides* cell? This process would be the equivalent of the step in nuclear transfer called enucleation in eukaryotic cells, which was relatively easy, since a micropipette could be used simply to suck out the nucleus from a recipient egg. We did not know if it was critical for our purposes to destroy or remove the genome in the recipient cell prior to the addition of the new DNA. Nor did we know what would happen if we ended up with two genomes in the same cell.

We thought of various ways to attack the host chromosome, including using radiation to destroy its DNA, with the reasoning that the much larger DNA molecules would be more affected by lower doses of radiation than the proteins. We also looked into the use of restriction enzymes to digest the DNA in the recipient cell genome. We remained concerned that any of these approaches might leave behind fragments of DNA, which might recombine with the transplanted chromosome, making it impossible to boot up a pure synthetic genome. After extensive discussion, Ham Smith proposed a simpler idea: do nothing, because after transplantation it was possible that, when the recipient cell divided into daughter cells, one of them might just end up with only the transplanted chromosome.

While this seemed like a long shot, we decided to conduct the experiment. But first we had to answer some key questions and develop a way to isolate and manipulate the chromosome without damaging the DNA. In order to protect the DNA, we chose to isolate the chromosome in tiny (100µl, or one-hundred-microliter) blocks of agarose, which has a consistency similar to that of gelatin. To start we placed the bacterial cells in a liquid agarose mixture and poured it into molds, which, when cooled in ice, solidified into tiny plugs. With the bacteria firmly contained, we could add enzymes to break the cells open so they would spill their contents, including the chromosomes, into the plug. To separate out the DNA we washed the plugs with protease enzymes to digest all the proteins, leaving the DNA intact. We could then place the plugs containing the DNA on top of an analytical gel and use an electrical current to drive it into the gel. DNA is negatively charged, due to the phosphate groups in the backbone, and will migrate toward a positive electrode when placed in an electrical field. By varying the voltage gradient and the gel percentage, and by using different stains, we could determine the size of the chromosome, the degree of protein contamination, and if the chromosome had been damaged or broken, to make them linear molecules, since linear DNA will move faster through the gels than circular DNA, which moves faster than supercoiled DNA.[12]

After experimenting with releasing and processing genomes into the analytical gels, we became confident that we had a way to isolate chromosomes, to determine if they were free of proteins (we needed to know if any proteins were required for transplantation), and to assess if they were double-stranded, circular, or supercoiled. In prokaryotes and eukaryotes, DNA is packaged in different ways to fit it compactly into cells or cellular compartments. In human cells DNA is wrapped around proteins, called histones, while bacteria tend to achieve the same feat by supercoiling, which, as the name suggests, means wrapping up coils within coils. Most bacterial genomes are "negatively supercoiled," meaning that the DNA is twisted in the opposite direction of the double helix. Some preliminary experiments had suggested that the precise state of the DNA was important, in that intact, circular chromosomes appeared to work best for transplantation.

Carole Lartigue and her team tried many approaches and ultimately settled on a procedure that, though complex, eventually worked. We learned that protein-free DNA would transplant successfully into *M. capricolum* cells. We also found that subtle changes in the recipient cells aided transplantation. For example, growing the cells to pH of 6.2 instead of 7.4 dramatically changed the appearance of the *M. capricolum* cells, usually ovoid in shape, making them appear long and thin. This slender topography also made them more permeable, presumably by making the cell membrane relax. To help the new genome get into the *M. capricolum* cells, we adopted a standard method that involved the use of a chemical called polyethylene glycol (PEG), which not only makes the membrane more permeable but may also protect the DNA as it moves across the membrane.

We found that the purity, type, and source of the polyethylene glycol were key factors in successful transplantation. The discovery of that simple fact took a lot of tedious, repetitive, and frustrating work, along with extreme attention to detail. When you are developing a new technique, there are no recipes to copy, textbooks to consult, or manuals to read to pass on those little tips and secrets that guarantee success. You end up having to try any and every permutation of conditions and ingredients. You are never quite sure which of the many factors is really significant, how they act with and against one another, and so on. To sort out all those variables requires carefully designed trials. This is basic experimentation at its toughest and, if you succeed, at its best. For every experiment that worked, there were probably hundreds that failed. It was a great credit to Carole Lartigue and her team that they worked out the details over many months of hard slog so that genome transplantation could be turned from an idea into a real, detailed, effective procedure.

In order to improve our chances of success, we added two gene cassettes to the *M. mycoides* genome prior to transplantation. One cassette was for antibiotic selection, so that when we added an antibiotic to the culture any cells that survived would have to carry this protective gene set. Also, to make it obvious when a genome transplant was successful, we included a gene called lacZ, which codes for a protein that would turn the recipient cells bright blue in the presence of a chemical called X-gal. Now we knew what success would look like: blue, antibiotic-resistant colonies. However, we had to be certain that was the case, since it was possible that blue could also result from the lacZ and the antibiotic-resistance genes being transferred to the *M. capricolum* genome.

Unfortunately, this was not just a theoretical possibility. We had tried numerous times to transplant the *M. genitalium* genome into *M. genitalium* cells, and although we frequently obtained blue cells, we found that it was always due to simple recombination events, in which the lacZ and resistance genes were indeed transferred from the nearly identical transplanted *M. genitalium* genome to the recipient cell's *M. genitalium* genome.

The transplanted genome was just too close in sequence structure to the recipient cell's genome, so their contents would mix. We learned the hard way not to get too excited at the sight of a blue colony before the achievement had been properly validated.

Once again, the team had some apparent success with getting new blue cells, following the transplantation of the *M. mycoides* genome into the *M. capricolum* cells. After obtaining one blue colony, they had refined and retested and, week by week and month by month, increased the numbers until we were getting dozens of colonies. Wiser now, we designed several experiments to analyze the blue cells that resulted from the genome transplantation.

Our first line of analysis used polymerase chain reaction (PCR) to amplify sequences that we knew occur only in the *M. mycoides* genome. Likewise, we had a number of sequences that occur only in the *M. capricolum* genome that we attempted to amplify from the blue cells. We would only start to get excited when we were able to detect amplified sequences from the transplanted genome and none from the recipient cells. There was still a lingering but small chance that we were seeing the results of genome recombination, but as we examined more and more sequences, that possibility became increasingly remote. We performed additional gel analysis of the transplanted cells, which revealed there were only *M. mycoides* fragments and none from the *M. capricolum* genome.

While these indirect methods were very encouraging, the ultimate test would be sequencing the DNA from a blue colony to reveal the actual genome content. We selected two of our blue colonies and sequenced thirteen hundred library clones from each, totaling over one million base pairs of DNA sequence. We became truly excited when all of the sequences matched only the *M. mycoides* genome that had been transplanted into the recipient cells.

With each stage of our analysis, it became clearer and clearer that we had cells that contained only the transplanted *M. mycoides* genome and that the *M. capricolum* genome had been destroyed or eliminated by segregation into daughter cells that were subsequently killed by the antibiotic in the growth medium. But we were still not satisfied. Could this finding be in some way an artifact? Was there any possible way that we could have transferred even one intact *M. mycoides* cell that grew and fooled us into thinking we had transplanted the genome? Were the blue colonies nothing more than the result of contamination? Nate Kaplan was the first to have taught me the old, wise mantra that extraordinary claims require extraordinary evidence to back them up.[13]

Guided by this critical thinking, we had included controls in every experiment that took us to this stage, so that we could rule out any artifacts. While we were certain that our DNA-isolation procedure would kill any and all *M. mycoides* cells, for reassurance we included two negative controls in each transplantation experiment: one in which the transplantations were done without recipient *M. capricolum* cells and one in which we had *M. capricolum* cells but the plugs contained no *M. mycoides* DNA. Using these negative controls, no blue colonies were ever observed, providing reassurance that our DNA preparations were not contaminated with any *M. mycoides* cells. We were further encouraged by the observation that the number of transplanted colonies from each experiment was dependent directly on the amount of *M. mycoides* DNA added to the cells. The more DNA we added, the greater the number of transplant colonies that resulted.

So what was it, exactly, that we now had? Was it *M. capricolum* cells that contained only the *M. mycoides* DNA, including the added lacZ and antibiotic-resistance tetracycline resistance tetM genes? What had changed in the wake of the genome transplant? What was the phenotype of cells derived from the transplanted DNA? We subjected the blue cells to a number of complex analytical procedures to find out what proteins were present. Using antibodies that were exquisitely sensitive to proteins in each parent-cell type, we investigated what the new transplant cells had on their cell surface. To our pleasant surprise the antibodies that were made against the *M. capricolum* proteins did not bind to the new cells with the transplanted genomes, whereas the antibodies that were made originally against the *M. mycoides* proteins did bind.

While these antibody studies were ongoing we also did a much more comprehensive analysis in which the proteins in all three cell types (recipient *M. capricolum* cells, donor *M. mycoides* cells, and new transplant cells) were examined by the method of differential two-dimensional electrophoresis. You can think of this technique as a method to view the protein content of cells. Proteins isolated from the cells are separated, in one dimension, based on their size, and in the second dimension, according to electric charge. This assay spreads out the proteins of the cell, each revealed as a pattern of spots that is unique for each cell type. The 2-D patterns can then easily be compared with one another. From this type of analysis it was clear that the transplant protein pattern was nearly identical to that from the donor *M. mycoides* cells and very different from the *M. capricolum* protein pattern.

We were thrilled by this result but wanted to go even further. We sequenced protein fragments from ninety different spots on the 2-D gels using a technique called matrix-assisted laser desorption ionization (MALDI) mass spectrometry. In this process, which would have sounded like something out of science fiction only a decade earlier, the tiny spots of separated proteins are, in effect, pulled out of the 2-D gels using a laser, forming a plume of charged molecules above each spot, and then subjected to analysis by a standard method called mass spectrometry. In this way, the MALDI procedure can reveal the amino-acid sequence of the protein fragments in the gel spots.

These data provided us with the conclusive evidence that the only proteins in the transplanted cell were those resulting from transcription and translation of the transplanted *M. mycoides* genome. We could now be absolutely confident that we had a new and novel mechanism to transform the genetic identity of a cell that does not involve DNA recombination or natural transformation mechanisms. Because the *M. capricolum* genome contained no code for processes to take up DNA, we could conclude that the transplantation of the new chromosome into the *M. capricolum* cells could only be the result of our polyethylene glycol genome-transplantation process. We now knew that we had the first cells derived from the deliberate transplantation of one species' genome into a host cell from another species. In doing so, we had effectively changed one species into another.

Our success had many implications. The most important was that we now knew that, if we could synthesize a genome from four bottles of chemicals, it was a realistic possibility that we could take that synthetic genome and transplant it into a recipient cell, where it would boot up its instructions. As a result, the transplant work reinvigorated our efforts to synthesize the DNA of an organism and then put it to work to create a new living cell.

The other major impact of the first genome transplants was that they provided a new, deeper understanding of life. My thinking about life had crystallized as we conducted this research. DNA was the software of life, and if we changed that software, we changed the species, and thus the hardware of the cell. This is precisely the result that those yearning for evidence of some vitalistic force feared would come out of good reductionist science, of trying to break down life, and what it meant to be alive, into basic functions and simple components. Our experiments did not leave much room to support the views of the vitalists or of those who want to believe that life depends on something more than a complex composite of chemical reactions.

These experiments left no doubt that life is an information system. I looked forward to the next goal. I wanted to put new information into life, to create a digital code on my computer, use chemical synthesis to turn that code into a DNA chromosome, and then transplant that man-made information into a cell. I wanted to take us into a new era of biology by generating a new life form that was described and driven only by DNA information that had been created in the laboratory. This would be the ultimate proof by synthesis.

8 Synthesis of the *M. mycoides* Genome

If we want to solve a problem that we have never solved before, we must leave the door to the unknown ajar.

—Richard Feynman, 1988[1]

Many believe that the most important innovations of human creativity are the result of some kind of visionary gift, a gift associated with such extraordinary and singular geniuses as Isaac Newton, Michelangelo, Marie Curie, and Albert Einstein. I don't doubt the incredible impact of individuals who can make great intellectual leaps, who can see further than anyone before them, and who discern patterns where others see only noise. However, there is also a less dramatic kind of creativity that drives science, a humble variety that is no less important: problem-solving.[2] Vaulting a single hurdle to achieve one very particular goal can sometimes result in a technology that can prove to have an extraordinary range of other uses. From a very narrow starting point science can be pushed in bold new directions.

When, for example, Hamilton Smith found out what a protein now called a restriction enzyme in the bacterium *Haemophilus influenzae* actually did, and how it did it, I doubt he had any idea that this discovery would help lay the foundation of genetic engineering. When the British geneticist Alec Jeffreys saw a fuzzy pattern on an X-ray film he'd made of genetic material from one of his technicians, he suspected that this messy image would pave the way to the science of forensic DNA—but I doubt he realized the extent to which it would come to be used routinely for paternity issues, for studying wildlife populations, and, of course, in criminal investigations. When Osamu Shimomura, a survivor of the Nagasaki atomic bomb, collected around a million individual crystal jellyfish (*Aequorea victoria*) between 1961 and 1988 to determine the secret of its bioluminescence (a protein called GFP), little did he realize that he would give the world a versatile, glowing tag that could provide vivid insights into how brain cells develop and cancer cells spread through tissue.[3] When we encountered a problem that held up our work on synthetic life—the genetic equivalent of figuring out how to run PC software on a Mac—our solution to it provided a bonus in the form of a new way to handle vast tracts of DNA.

By that time we had carried out successful genome transplantations in 2007 and also completed the laborious assembly of laboratory chemicals into a 582,970-base-pair *M. genitalium* genome. We had created a synthetic bacterial chromosome and had successfully grown it in yeast. We had also been able to purify the synthetic chromosome out of yeast so that it could be checked by DNA sequencing. However, our process of using yeast cells as the vehicle for the final assembly of the chemically synthesized genome segments meant that we had our prokaryotic (bacterial) chromosome growing in a eukaryotic (yeast) cell. In order to complete the construction of our synthetic cell, we needed to isolate the synthetic chromosome from yeast in a form that would allow it to be transplanted into the recipient prokaryotic cell.

At this point we ran into a number of problems that we had not anticipated. The first concerned the conformation of our synthetic chromosome. We had been able to purify the synthetic chromosome in the linear and circular forms for DNA sequencing but had found evidence in our detailed work on genome transplantation indicating that the chromosome may need to be intact, which means that it cannot have any cuts or nicks in the DNA. Our purification method was too harsh to give us intact DNA. We had effectively created a digital recording in a format that could not be read by any player.

The second problem was that our effort to extend our genome-transplant work to another mycoplasma species, *M. genitalium*, had been unsuccessful. When we transplanted the *M. genitalium* genome back into *M. genitalium* cells, the chromosome always recombined with the existing genome. That was not the only problem. As we already knew only too well, although *M. genitalium* had the smallest genome, it was not an ideal organism for the development of new methods, because of its frustratingly slow rate of growth. Each experiment took up to six weeks before colonies would appear. Nonetheless, we struggled on. While these experiments continued, we decided that, due to our success with transplanting the *M. mycoides* genome into *M. capricolum* cells, we would use these faster-growing species to solve the problem of obtaining intact chromosomes out of yeast cells. Our first step was to see if we could clone the entire *M. mycoides* genome in yeast. This target seemed achievable, given our success in creating a synthetic genome in an artificial yeast chromosome.

This challenge was given to Gwyn Benders, a postdoctoral fellow working with Clyde Hutchison. Large DNA molecules were by then routinely and stably grown in yeast by the addition of a yeast centromere, a specialized region of the genome, which can be identified under the microscope as the constriction in chromosomes when they form a characteristic X-shape during cell division. This pinched region is the nexus of a chromosome, which plays an essential role in ensuring that, when a cell divides, each daughter inherits a copy of every chromosome. Thus, when a centromere is added to a large piece of DNA, the latter can be copied and segregated, along with the yeast chromosomes, during cell division. With this method, we could grow circular chromosomes. By adding telomeres to them—structures found at the end of chromosomes—the foreign (exogenous) DNA molecules could be grown in a linear form as well.

Building on this earlier work, we developed three different approaches to cloning whole bacterial chromosomes in yeast. In the first, we chose to insert the artificial yeast centromere into the bacterial chromosome prior to its being grown in yeast. In the second and third approaches, we did everything simultaneously: introducing the bacterial chromosome and the yeast centromere with overlapping sequences that would result in their recombination within yeast. Encouragingly, all three approaches were successful, and in a range of genomes, including the *M. mycoides* genome, *H. influenzae* genome, and a photosynthetic algae genome.[4] It is very interesting that the only requirement to convert a bacterial chromosome into a yeast chromosome is the addition of a small synthetic yeast centromere together with a selectable marker. We now had a method that we could use to test several different ways to make a genome transplantation work. Once we had the *M. mycoides* genome stably cloned into yeast cells, we developed procedures to recover intact chromosomes in a transplantable form.

To test the yeast-derived *M. mycoides*, we transplanted it as before into the *M. capricolum* cells. However, even though this experiment was attempted numerous times, the team was unable to recover any transplant cells. They ran controls using the *M. mycoides* genome isolated from wild-type *M. mycoides* cells, and these genome-transplant experiments always worked, as they had before. As I waited for the results, Ham Smith came to see me and delivered his devastating verdict: "Genome transplants out of yeast do not work." The yeast that had allowed us to manipulate large stretches of bacterial DNA seemed to be the source of the problem.

Ham Smith and I had earlier discussed what might distinguish an *M. mycoides* genome grown in *M. mycoides* cells from the same genome grown in yeast cells, so during a video conference between the Rockville team and the team in La Jolla we quickly focused on the most obvious difference. Within the *M. mycoides* cells the DNA is specifically methylated (decorated with molecular tags called methyl groups). These groups, derived from methane, each consist of one carbon atom bonded to three hydrogen atoms (CH_3). Bacterial cells use DNA methylation to protect the DNA from being cut by its own restriction enzymes—by adding methyl groups to the DNA bases recognized by the restriction enzyme. Yeast, however, does not have the same restriction enzymes and DNA methylation systems, and the bacterial methylases were unlikely to be expressed in yeast because of the codon differences. We reasoned that if the *M. mycoides* genome in yeast was indeed unmethylated, on transplant into the *M. capricolum* the restriction enzymes in the host cells would immediately destroy the transplanted genome.

We set out to try two approaches to test the idea that the lack of DNA methylation in yeast was the reason for the transplant failures. In the first we cloned the six *M. mycoides* DNA methylation genes to produce enzymes to methylate the *M. mycoides* genome after its isolation from yeast cells. This method was successful; the genome transplantation finally worked with the methylated genome isolated from yeast cells. If we methylated the *M. mycoides* genome with *M. mycoides* cell extracts or with cloned and purified DNA methylases, we could likewise successfully transplant the genomes into the *M. capricolum* cells. As the final proof that it was the DNA methylation that was the relevant factor, we removed the restriction-enzyme gene from the recipient *M. capricolum* genome, reasoning that, if the recipient cell lacked restriction enzymes, then we should be able to transplant naked *M. mycoides* DNA from yeast directly, without the need for protective methylation. This was indeed the case, and at long last we seemed to have all the required components to transplant synthetic chromosomes.

The brief synopsis above does not reflect the fact that solving the methylation problem actually required over two years of toil, but in the process we had developed a very powerful new set of tools that enabled us to manipulate bacterial chromosomes in a way never before possible. Most types of bacterial cells do not have genetic systems (such as homologous recombination) that are found in yeast and *E. coli*. As a result it is extremely difficult, if not impossible, to make substantial genetic changes to the majority of bacterial genomes. This is one of the reasons most scientists work with *E. coli*: they do so simply because they can.

Now, however, by adding a eukaryote (yeast) centromere to a prokaryote (bacterial) genome, we were able to grow the bacterial genome in yeast, where it behaved just like a yeast chromosome. That paved the way for the use of the yeast homologous recombination system to make numerous and rapid changes in the genome. We were then able to isolate the modified bacterial genome, methylate it if necessary, and transplant it into a recipient cell to create a new cell version. That marked a major advance in genetic manipulation. Before this discovery, scientists had largely been limited to tinkering with individual genes. Now we could routinely manipulate sets of genes and even entire genomes. We published the results in *Science* in September 2009.[5]

By now we had developed new ways to synthesize DNA at a scale twenty times what had been possible before; we had developed the methodology to transplant the genome from one species to another to create a new species; and we had solved the DNA-modification problems of restriction enzymes destroying transplanted DNA. We and many of those scientists who were following our progress were ready to see if we could finally succeed in creating a cell based on a synthetic genome. Now that we had shown that we could go from yeast to mycoplasma, the next step was going from synthetic DNA in yeast to mycoplasma. We were ready to combine all of these novel approaches to make history. But were we using the right mycoplasma?

As we were making great progress in using yeast to manipulate large pieces of DNA, our teams doggedly continued with their attempts to transplant the synthetic *M. genitalium* genome, but without much success. It appeared as though we could convert the larger *M. pneumonia* species into *M. genitalium* by transplantation of the native genome, but we could not achieve the same result with the synthetic version. We discovered, eventually, that the recipient *M. pneumonia* cells contained a nuclease on their cell surfaces that chewed up any DNA to which it was exposed.

But we continued to be plagued by the extremely slow growth of the *genitalium* cells, which greatly limited the number of experiments that could be done. To all of us, but particularly to me, having to wait weeks for results was not just frustrating; it was painful. Something had to change. While we were making advances with DNA methylation and transplantation, we had also figured out a quicker way to create synthetic DNA, a technical feat that would provide new opportunities in our quest to transplant synthetic bacterial DNA. As I described earlier, our initial stages of DNA synthesis required several steps, starting with assembling short oligonucleotides into larger constructs. However, Dan Gibson had greatly simplified this method, so that we now had a single-step process.

The DNA-assembly reactions that Dan had used for his new method were very similar to those we had employed in our earlier work, except that he made the important discovery that they could all be carried out in one tube and at a single temperature. Dan had realized that the exonucleases that cut back one strand of DNA do not compete with the DNA polymerase that we used to fill in missing bases of DNA. Also, when all the enzymes are combined in a single reaction at fifty degrees centigrade, the exonuclease is rapidly inactivated at this temperature and therefore can chew back only enough bases to allow the DNA fragments to anneal with each other.

This work represented a great advance over our previous approach, which was tedious and labor-intensive. The power of this "Gibson assembly," as we call it, lies in its great simplicity. Until then, we had been reluctant to contemplate a repeat of the marathon effort it had taken to create our first synthetic chromosome, *M. genitalium* JCVI-1.0, from ten thousand pieces of DNA, let alone attempt anything larger. From the beginning we believed that the chemistry of genome synthesis would be the toughest problem to solve; now the development of the synthesis technology gave me the confidence to make a significant change in direction for the program.

I spoke with Ham Smith and said that we had been taking the wrong approach. Because we would likely never be successful working—more like struggling—with slow-growing *M. genitalium*, I told him, "I want us to stop all work on it and use our new techniques to synthesize the *M. mycoides* genome." That was a tall order, since the *mycoides* genome was twice as large as that of *genitalium*. But by that time we knew how to transplant the *M. mycoides* genome from yeast, and how to make *M. mycoides* cells out of *M. capricolum* cells; armed with the Gibson assembly, we should in theory have been capable of constructing a 1.1-million-base-pair genome relatively quickly.

Initially I encountered strong resistance from my team, and in retrospect I'm not surprised by the immediate pushback to this radical proposal to not only start all over again but to aim for an even more ambitious target. Dan Gibson, not surprisingly, agreed with my change of plan, while Ham Smith and the rest of the team wanted to stay on the existing track, with its familiar set of problems, but focus on finding ways to solve them, even if it did mean grinding on with the slow-growing *genitalium* cells. After several rounds of discussion, however, and after everyone had had plenty of time to mull my idea over, minds began to change. Ham came to see me and agreed that we should change directions; we immediately called Dan Gibson and told him to get started on the synthesis of the *M. mycoides* genome.

One of the reasons that the team had initially been resistant to this new approach was that we did not yet have an accurate genome sequence of the *M. mycoides* genome. We started the design and synthesis while we quickly sequenced two strains. Our two genome sequences from *M. mycoides* isolates differed from each other at ninety-five sequence sites. Because yeast now played a central role in assembling synthetic genomes, we chose the sequence from the isolate that we had already successfully cloned in yeast and transplanted into the recipient cells.

We began by designing 1,078 cassettes, each of which was 1,080 base pairs in length and overlapped with its neighbors by eighty base pairs. To enable us to cut out the assembled sequences from the vectors they were grown in, we added to the cassettes another eight-base sequence recognized as cutting sites by the NotI restriction enzyme. We also included four watermarks to help distinguish our synthetic genome from any naturally occurring species. The watermarks ranged from 1,081 to 1,246 base pairs and contained a unique code that was designed to enable us to write numbers and words in English. We took care to insert the watermark sequences into genome regions that we had shown experimentally would not affect the cell's viability. With the design complete, we ordered the 1,078 cassettes from Blue Heron, an early DNA-synthesis company, based in Bothell, Washington. The company assembled the 1,080 base-pair segments from chemically synthesized oligonucleotides, sequenced them to make sure they met our specifications, and shipped them to us.

Now we could start the construction of the synthetic organism in earnest. We used a hierarchical construction that took place in three stages. First, using the Gibson assembly, we built from the 1,080-base-pair cassettes a series of 111 longer cassettes, each around ten thousand base pairs (10kb). As ever, accuracy was paramount, so we sequenced all 111 of them and found errors in nineteen. These sequence errors were corrected, and the 10kb clones were then reassembled and resequenced to verify that they were correct.

In the second round of the genome assembly, we assembled overlapping 10kb cassettes to form 100,000-base-pair cassettes (which we also sequenced to check our accuracy). Finally, we combined the resulting eleven 100kb cassettes together in yeast to obtain the 1.1-million-base-pair sequence of the *M. mycoides* genome. Once again, we had to verify that the assembly had worked, and used PCR and restriction-enzyme digestion to confirm that we had the correct genome structure.

At last we were ready to attempt to create the first synthetic cell by transplanting the fully synthetic *M. mycoides* genome out of yeast into the recipient *M. capricolum* cells. As before, success would be signaled with blue cells. The first transplantation was done on a Friday, and we anxiously waited over the weekend to see if any blue clones would appear by Monday morning. But Monday came and went with no positive results. The transplantation was repeated the following two Fridays, but again with no blue colonies, only blue Mondays.

In retrospect it's clear that we were very close to success, though it did not feel so at the time. After running all the controls we were convinced that a design or sequence error must have managed to slip past the various checks we had made during genome synthesis. Because we had sequenced the DNA we assumed that the error must have occurred in one of the original sequences that we had used in the genome design. In order to check our code we had to create the biological equivalent of what the developer of a computer application would recognize as debugging software. The operating system of a modern computer is vast, running to tens of millions of lines of code,[6] and over the decades engineers have developed smart debugging programs to aid in finding faults.

Vladimir Noskov, a staff scientist in the Synthetic Biology & Bioenergy Group at the JCVI, Maryland, was our resident yeast guru. Noskov had graduated from St. Petersburg State University, Russia, and then went on to do a doctorate there in yeast genetics. After spending five years in Japan studying chromosomal DNA replication and a "checkpoint" in the yeast cell cycle, where DNA surveillance and repair is carried out, he worked at the National Institutes of Health, Bethesda. There, in the Chromosome Structure and Function Group, Noskov invented several applications for a technique to manipulate large pieces of DNA in yeast, known as transformation-associated recombination (TAR) cloning, which offers advantages over an older method that relies on what are called yeast artificial chromosomes, or YACs.

For our biological debugger we decided to start by verifying the eleven 100,000-base-pair segments. Noskov, using TAR cloning, constructed equivalent-size segments from the native *M. mycoides* genome so that we could independently substitute each one with a synthetic segment to see if they would support life. From these complex experiments we found that all but one of the eleven 100kb synthetic segments was compatible with life. For final proof Dan Gibson constructed a hybrid genome with ten synthetic segments and one native segment and obtained successful transplants.

Having established which segment contained an error or errors that did not support life, we sequenced the DNA once again, this time using the highly accurate Sanger sequencing method, and found that there was a single base-pair deletion. If this sounds as trivial as writing "mistke" instead of "mistake," equating nucleotides to individual letters is slightly misleading, in the sense that DNA code is read three nucleotides at a time, so that each three-base combination, or codon, corresponds to a single amino acid in a protein. This means that a single base deletion effectively shifts the rest of a genetic sentence that follows, and hence the sequence of amino acids that the sentence codes for. This is called a "frameshift mutation"; in this case, the frameshift occurred in the essential gene dnaA, which promotes the unwinding of DNA at the replication origin so that replication can begin, allowing a new genome to be made. That single base deletion prevented cell division and thus made life impossible. Once we had found the critical error, we were able to reassemble the 100kb segment correctly and used yeast to reassemble the entire genome. We were now ready once again to attempt the genome-transplantation experiment.

As before, the critical experiment was initiated on a Friday, so that any successful colonies would have enough time to grow to appear as blue dots on the following Monday. Dan Gibson reported on the latest attempt in an e-mail to us:

> Craig, Ham, Clyde, and John, the completely synthetic genome (with four watermarks and no dnaA mutation) is being transplanted today. The genome looks great. It was analyzed by multiplex PCR at each of the 11 junctions and at the four watermark sequences. It was also checked by restriction digestion and FIGE analysis. At the same time, two yeast clones containing 10/11 semi-synthetic genomes are being transplanted. These genomes were analyzed as above and also look great. I will send an e-mail on Monday morning but keep in mind that colonies have been appearing later than usual so we may not have an answer until Tuesday.

That afternoon, Dan had handed a tiny vial to his colleague Li Ma, who was sitting before a bio-safety hood—one of the enclosed, HEPA-filtered workspaces that we use in our laboratories to work under sterile conditions. The vial contained a tiny plug of agarose in which were embedded a few million microscopic circular DNA chromosomes, each corresponding to our synthetic genome of 1,078,809 base pairs. This was the synthetic DNA code for the 886 genes of *M. mycoides*, along with our watermarks. Li added drops of the enzyme that would dissolve the gel to leave behind the synthetic genomes, which she then added to a second tiny vial that contained the recipient *M. capricolum* cells, followed by polyethylene glycol to make the membranes permeable to the DNA. Li spread the cells on a Petri dish of red agar, a recipe that would nourish the new cells with sugar and amino acids. The agar was also laced with tetracycline to kill any remaining recipient cells that did not take up the synthetic genome and X-gal that the new cells would use to turn bright blue if they contained the transplanted genome. Late that afternoon Li placed the Petri dishes in an incubator to keep the cells at constant thirty-seven degrees centigrade. If any new cells were produced, they would need a couple of days to divide enough times to produce one million daughter cells, a small colony visible to the naked eye.

Over the weekend my anxiety level was very high. The moment to see whether our genome modifications had been successful seemed to take forever to arrive. Finally, very early on Monday morning, Dan Gibson, his hands shaking with anticipation, opened the door of the incubator and began to remove the Petri dishes one by one. Leaving the best 'til last, he began with the control plates (the ones that would indicate that Li Ma had followed the correct procedures) and held each up to the light to check if any colonies were visible. He then moved on to the ones containing synthetic DNA. There, a little off the center of one plate, was one—and only one—bright-blue colony of cells.

Dan paused to appreciate the significance of that moment. After staring at the Petri dish for a few minutes, he placed all the plates back in the incubator and sent me a simple text message at 4 A.M. Rockville time: "We have a blue transplant." We had come such a long way, suffered so many failures, so many years of efforts of trial and error, of problem-solving and invention. At last, it seemed that all the effort had finally paid off.

While the experiment was playing out in Rockville I was staying at our townhouse in Alexandria, Virginia, so I was in the same time zone as Dan. A series of exchanges soon followed. Dan e-mailed again: "So far, one or two blue colonies have arisen from the transplantation with the complete synthetic genome! It is too early to get an accurate count of the number of transplants but I am sure that this number will be higher. We'll look again later today and do a final count tomorrow." We needed to validate that the blue colony contained only the synthetic DNA. I told him that I would be at the institute by 10 A.M. and asked how long it would take for the first validation results.

I took my video camera and a couple of bottles of champagne with me as I headed down the George Washington Memorial Parkway toward Rockville. As soon as I arrived, I went directly to the transplant room, where I encountered Dan, beaming and already surrounded by the rest of the team. Following some warm handshakes, Dan took me over to the incubator and pulled out the culture dish to show me the first blue colony. We took some photos of it and carefully placed what was becoming very likely to be the first life form with a completely synthetic genome back into the incubator.

Later that day I received the first confirmation that the culture dish contained the synthetic genome: "Congratulations! It's official. You are the proud parent of a synthetic *mycoides* cell!" I gathered the entire team in the boardroom in Rockville and video-linked us to the rest of the synthetic-genome team, in La Jolla. (I had made sure that the La Jolla team also had plenty of champagne on ice.) Even though the final verification of the results would take a few days, we all happily toasted our apparent success.

The first confirmation came from Dan at 7:45 on Tuesday morning: "Good news. All 4-watermark sequences show up in multiplex PCR in the single synthetic transplant. The watermarks do not show up in the WT [wild-type] and *M. capricolum* negative controls which have no colonies." On Thursday, April 1, Dan e-mailed the results of a second round of experiments: "The complete synthetic genome was transplanted again. This time it produced a lot of colonies! In addition, a second synthetic genome clone also produced a lot of colonies. I will now move the Happy Birthday balloon into the transplant room."

The very next day we had even stronger confirmation that it was only the synthetic genome controlling the new cells: "Great news! The synthetic transplant produces the expected restriction fragments when digested with AscI and BssHII." (Sites that were cut by these restriction enzymes had been added to three of the four watermark sequences.) On April 21 we had the results from sequencing the DNA from the living synthetic cells, and there was no longer any room for doubt: the cell had been controlled only by the genome that we had designed and synthesized. The sequence showed our genome to have the 1,077,947 base pairs, exactly as intended, including nineteen expected differences from the native genome, as well as the four watermark sequences, a critical part of the proof that the DNA was synthetic. Just as we had suspected, a one-letter deletion out of over one million base pairs of DNA had made the difference between life and no life. I can think of no more dramatic illustration of how information plays a central role in life.

We had put a significant effort into the design of our watermarks to ensure that we could safely code for complex messages in the DNA sequence. In the first synthetic genome we used the single-letter abbreviation of the amino acid coded by the triplet codon to represent letters of the English alphabet. For example, the triplet ATG codes for the amino acid methionine, which uses the letter M as an abbreviation. But because the range of amino-acid codes doesn't include abbreviations that would cover the 26-letter alphabet, we designed a much more complete system that would enable us to code the entire English alphabet together with punctuation, numbers, and symbols. (The series ABCDEFGHIJKLMNOPQRSTUVWXYZ [NEWLINE] [SPACE] 0123456789#@)(-+\=/: <;>$&}{*]%!'., was represented, respectively, by TAG, AGT, TTT, ATT, TAA, GGC, TAC, TCA, CTG, GTT, GCA, AAC, CAA, TGC, CGT, ACA, TTA, CTA, GCT, TGA, TCC, TTG, GTC, GGT, CAT, TGG, GGG, ATA, TCT, CTT, ACT, AAT, AGA, GCG, GCC, TAT, CGC, GTA, TTC, TCG, CCG, GAC, CCC, CCT, CTC, CCA, CAC, CAG, CGG, TGT, AGC, ATC, ACC, AAG, AAA, ATG, AGG, GGA, ACG, GAT, GAG, GAA, CGA, GTG.) This cipher was the key to decoding the watermarks. The first watermark included "J. Craig Venter Institute" and "Synthetic Genomics Inc.," the names of several scientists, and the message "Prove you've decoded this watermark by emailing us (mail to: MROQSTIZ@JCVI.org)." As the first living self-replicating species to have a computer as a parent, it had to have its own e-mail address.

With our first synthetic genome we were limited by using the amino-acid code to add any significant messages; now, with our new code, I wanted to mark the historic moment by including some apposite quotations from the literature. I found three that I felt were both significant and relevant to the first synthetic life form. The first quotation is to be found in the second watermark: "To live, to err, to fall, to triumph, to recreate life out of life." That, of course, comes from James Joyce's *A Portrait of the Artist as a Young Man*. The third watermark included, alongside the names of several scientists, the second quotation: "See things not as they are, but as they might be," attributed to one of Manhattan Project physicist J. Robert Oppenheimer's early teachers and cited in the biography *American Prometheus*. The fourth watermark contained the remaining names of the forty-six scientists and a quotation from the Nobel Prize–winning quantum physicist Richard Feynman: "What I cannot create, I do not understand."

We had accomplished what had been only a wild dream almost fifteen years earlier, and had effectively come full circle. Starting with DNA from cells, we learned how to accurately read the DNA sequence. We successfully digitized biology by converting the four-letter chemical analog code (A, T, C, G) into the digital code of the computer (1s and 0s). Now we had successfully gone the other direction, starting with the digital code in the computer and re-creating the chemical information of the DNA molecule, then in turn creating living cells that, unlike any before, had no natural history.

The assumption, at least by most molecular biologists, was that DNA and the genome, represented by the sequence of letters in the computer, was the information system of life. Now we had closed the loop by starting with the digital information in a computer and, by using only that information, chemically synthesized and assembled an entire bacterial genome, which was transplanted into a recipient cell, resulting in a new cell controlled only by the synthetic genome. We named our new cell *M. mycoides* JCVI-syn 1.0 and worked on getting our results ready for publication.

I submitted our manuscript, with twenty-four coauthors, to *Science* on April 9, 2010. We briefed White House officials, members of Congress, and officials from several government agencies before the paper was accepted for publication the following month, on May 13.

On the day of its publication online, May 20, 2010 (the print version appeared July 2),[7] media from around the world gathered in Washington, D.C., for our press conference. Joined by editors from *Science*, we announced the first functioning synthetic genome. Ham Smith explained to the gathering that we now had the means to dissect the instructions of a cell to determine how it really worked. We also discussed our larger vision—namely, that the knowledge gained in doing this work would one day undoubtedly lead to a positive outcome for society through the development of many important applications and products, including biofuels, pharmaceuticals, clean water, and food products. When we made the announcement, we had in fact already started working on ways to produce vaccines and create synthetic algae to turn carbon dioxide into fuel.

9 Inside a Synthetic Cell

The first pillar of life is a Program. By program I mean an organized plan that describes both the ingredients themselves and the kinetics of the interactions among ingredients as the living system persists through time.

—Daniel E. Koshland, Jr., 2002[1]

Definitions are important in science. But sometimes it is just as important not to get too obsessed by them, particularly when you are venturing into a new area, as they can be distractions that hinder how you think and what you do. They can become a trap, as they did in the first half of the twentieth century, when scientists were certain that proteins were the genetic material. Richard Feynman issued a famous warning about the dangers of attempting to define anything with total precision: "We get into that paralysis of thought that comes to philosophers . . . one saying to the other: 'You don't know what you are talking about!' The second one says: 'What do you mean by "talking"? What do you mean by "you"? What do you mean by "know"?'"[2]

When we unveiled the details of the first synthetic organism in our *Science* publication, we defined what we had done and how we had done it. We defined the terms "synthetic life" and "synthetic cells" in a reasonably specific way, as cells completely controlled only by a synthetic DNA chromosome. The synthetic genome was the software of life, which specified every protein robot in the cell, and therefore every one of the cell's functions. From the public response to our announcement and scientific publication, however, it was clear that some found it hard to accept the concept of life as an information system.

This skepticism became even more evident in the subsequent press coverage from around the world. Most of the accounts were very positive—or even *too* positive. One professor declared that I was "creaking open the most profound door in human history" and added that I was "going toward the role of a god."[3] Some of the reports were more sober and qualified. The BBC declared that it was a "breakthrough," a somewhat overused word, and *Time* listed it among its top medical advances of 2010. *The New York Times* cited researchers who thought we had achieved a technical tour de force, if not a true breakthrough. Biologists were in general impressed, according to *Maverick Genius*, Phillip F. Schewe's biography of Freeman Dyson, the influential English-born professor of physics at the Institute for Advanced Study, in Princeton. Dyson himself was quoted as saying that my experiment was clumsy but "important work, as it was a big step toward creating new forms of life."[4] Then, inevitably, came the knee-jerk concerns of some extreme environmentalists[5] and the customary sensationalism of the British tabloids.[6] One had asked plaintively of our cell, "Could it wipe out humanity?"[7]

The most significant criticisms focused on the true significance of creating a cell controlled by DNA software. Did it count as synthetic life? Some pointed out correctly that our synthetic genome was closely based on an existing genome and thus did not count as being truly synthetic, having a natural ancestor in the form of *M. mycoides*. But, as Schewe remarked, there were also those biologists who were absolutely certain that we had not created synthetic life at all because we had used a natural recipient cell, arguing that this term should be reserved for the creation of a living thing "from scratch." And, indeed, President Obama's Bioethics Commission agreed that our work, "though extraordinary in many ways," and a proof of principle, did not amount to creating life, as we had used an existing natural host, an already-living cell.[8] A number of milder versions of this argument made various different points in an attempt to downplay the significance of what we had achieved. The Vatican newspaper *L'Osservatore Romano*,[9] in an otherwise mostly positive and helpful statement, concluded that our team had not created life but had only "changed one of life's motors."

This diverse range of views tells us several things. There is still no agreed-upon definition of what we actually mean by that troublesome word "life," let alone "synthetic life," "artificial life," or "life from scratch." The definitions were of course dependent on the traditions of those who had crafted them. The term "artificial life" had a quite different meaning in scientific circles in the 1990s, when it was mostly used to refer to replication in a computer. One example can be found in the work of Thomas S. Ray, who in 1996 wrote about systems that "evolve freely within the digital medium, like the evolution by natural selection in the carbon medium that generated life on Earth." The primary objective of this work, he explained, "is to provoke digital evolution to generate complexity within the digital medium, comparable in magnitude to the complexity of organic life."[10]

There is a sharp distinction between our work with synthetic chromosomes in wet biology and the simulation of artificial life in hard silicon. "Artificial life" is a term traditionally used to describe that which occurs in the digital world, whereas "synthetic life" has its origins in the digital world but also encompasses life in the biological world. Even so, life in vivo and in silico are both united by the concept of information-driven systems, and our synthetic-cell work provides the first direct link between the two.

We now know that the right DNA code, presented in the right order and placed in the right chemical context, can produce new life out of existing life. With the synthetic cell, we built upon 3.5 billion years of evolution, and did not try to recapitulate it: because we had modified the genome, there was no direct ancestor of the cell we had created to be found in nature. With our synthetic code we had added a new tributary to the river of life.

We now know how to write that code *de novo*, with the help of computers, which opens up the potential for designing almost any type of living thing, as we discover more details of the machinery of life. In the wake of this work in my laboratory, we can define "synthetic life" as self-replicating biology based on a synthetic genome, a synthetic code-script. As I write this, my teams have designed our first attempts at a minimal genome, based mostly on first principles, consisting of those genes we think are necessary for life. As discussed above, the genome still contains a significant percentage of genes whose functions have not been determined, other than that, through detailed experiments, we know that they are absolutely required for the cells to live. We are using a recipient cell, just as we did with the first synthetic cell, to boot up this new software of life.

The implications of our ability to design life are profound. Since the pioneering work of Robert Hooke, in the seventeenth century, we have known that all living things are constructed from one or more cells. Today, by adjusting their genetic programming, we can potentially alter the structure and function of any cell we like, to create an amazing variety of life, from undersize yeast cells[11] to fast-growing fish.[12] We can also explore the ancient mechanisms used to generate the three-dimensional organization of a cell from one-dimensional genetic software.

So far as we know, all the cellular life that exists on our planet originated from earlier kinds of cells. Every single one of these fundamental units of life, including the Earth's 5,000,000,000,000,000,000,000,000,000,000,000 or so bacteria,[13] is descended from the first cells that lived around four billion years ago. Whether or not these cells came from another planet, through the process known as panspermia, or were spread by intelligent life forms via what Francis Crick called "directed panspermia,"[14] the ultimate origin of the first cells remains a mystery.

When there is mystery, there is an opportunity for vitalism and religion to thrive. However, when my team successfully booted up synthetic DNA software within a cell, we demonstrated that our basic understanding of the machinery of cellular life had advanced to a significant point. In answer to Erwin Schrödinger's little question "What is life?" we had been able to provide one compelling answer: "DNA is the software and the basis of all life."

Yet because we began with an existing cell and all its protein machinery, the question remains whether modern cells, which are the result of billions of years of evolution, can actually be recreated from the basic components of life. Can we induce all the complex cellular functions to operate without initially needing the protection of a cell membrane, and, if so, can we utilize the isolated protein and chemical components to boot up a synthetic chromosome and in the process create a new kind of self-replicating cell? Can we grow in the laboratory an organism that represents a brand-new branch on the tree of life, a representative of what some like to call the Synthetic Kingdom? In theory, at least, we can. The science of the coming century will be defined by our ability to create synthetic cells and to manipulate life.

My confidence is based in part on the enormous advances that have been made since 1965, when it was first suggested that the synthesis of living cells become a national goal for America.[15] Over the past few years we have seen the rise of synthetic biology, an emerging phase of research in molecular biology. The field represents a marked shift away from the reductionist experimentation that, over the decades, has been a powerful method to help us understand cells, by revealing their constituents, dynamics, and cycles. We now have to see if we can assemble all those myriad cellular components in novel ways to create life anew. When we achieve that milestone, we will have opened a new chapter in our understanding of life and, I believe, will have a complete answer to Schrödinger's difficult question.

Even when we achieve life from a cell-free system, it still cannot be considered "life from scratch," whatever that might mean. I doubt if any of the individuals who have used this phrase have thought much about what they are actually trying to express with it. Let's use baking a cake "from scratch" to illustrate what I mean. One could imagine buying a cake and then icing it at home. Or buying a cake mix, to which you add only eggs, water, and oil. Most consider baking a cake "from scratch" to involve combining the individual ingredients, such as baking powder, sugar, salt, eggs, milk, shortening, and so on. Using "scratch" in this context, I doubt that anyone would mean formulating his own baking powder by combining sodium, hydrogen, carbon, and oxygen to produce sodium bicarbonate, or producing homemade cornstarch, which is a highly branched carbohydrate comprised of a large number of glucose units joined by glycosidic bonds. Glucose, for its part, is formed from carbon, hydrogen, and oxygen. If we apply the same strictures to creating life "from scratch," it could mean producing all the necessary molecules, proteins, lipids, organelles, DNA, and so forth from basic chemicals or perhaps even from the fundamental elements carbon, hydrogen, oxygen, nitrogen, phosphate, iron, and so on. The origin of the primary ingredients themselves, organic chemicals, is missing the point, though it *is* relevant to the huge question of where life came from in the first place. The nature of chemistry at the dawn of life—prebiotic chemistry—takes us back to 1952 and the famous experiments by Stanley Miller and Harold Urey, at the University of Chicago. Complex organic molecules, including sugar and amino acids, were formed spontaneously from water (H_2O), ammonia (NH_3), hydrogen (H_2), and methane (CH_4) when the latter were exposed to conditions (a closed, sterile system with heat and sparks from electrodes) that simulated those thought to have been present on the early Earth.[16] A few years later, at the University of Houston, Joan Oró found that the nucleotide base adenine and other RNA and DNA nucleobases could form spontaneously from water, hydrogen cyanide (HCN), and ammonia.[17]

Many believe that RNA was the first important replicating genetic material, the precursor of DNA-based life, and describe an "RNA world."[18] In 1967 Carl Woese was one of the first to suggest that RNA might have catalytic properties, so that it both carried genetic information (like DNA) and could behave like proteins (enzymes) as well, which is significant because virtually all chemical reactions that take place within a living cell require catalysts.[19] It was not until 1982, when Thomas Cech, at the University of Colorado, Boulder, showed that an RNA molecule could self-splice out an intron,[20] together with the discovery by Sidney Altman, at Yale University, of the catalytic properties of ribonuclease P,[21] which can cut RNA, that we knew for sure that these catalytic RNAs—"ribozymes"—actually existed. Cech and Altman shared the 1989 Nobel Prize in Chemistry for their discoveries.[22]

Ribozymes may be key in trying to answering the most basic question of all. How did the very first cell evolve, whether here on Earth or on an exoplanet?[23] Many approaches have been taken to try to understand the origins of life, but if there is one researcher who has attempted to do so by actually forming primitive "life" (from scratch), it is Nobel laureate Jack W. Szostak,[24] along with his lab, at Harvard University. Unlike teams working on "artificial cells" that are composed of protein systems in lipid vesicles but lack any of life's software molecules, Szostak recognizes that life requires a self-replicating "informational genome."[25] Szostak's view falls between two camps engaged in studying the origins of life. There is the software-first camp, which argues for the emergence of RNA replication, as both an information-carrier and a catalytic molecule, as the most important step in the onset of life. The other group argues that the key factor in the evolution of the first life was the emergence of the cell membrane, in the form of self-assembling and self-replicating vesicles.

These vesicles, bubble-like structures that are also known as micelles, are formed spontaneously from lipid molecules when those molecules are present above certain concentrations. The earliest lipid molecules are thought to have been fatty acids, which were present on the early prebiotic Earth and have even been found in meteorites. At the molecular level, they have a hydrophobic (greasy, water-hating) end and a hydrophilic (water-liking) end, which can link up to form structures. The lipid molecules join tail to tail (greasy end to greasy end), leaving the water-liking ends exposed on the inner and outer cell-membrane surfaces. This assembly acts as an effective barrier to keep water-soluble molecules contained inside the cell to create a unique environment.

Working on experiments with his student Irene Chen, together with Richard J. Roberts,[26] at Caltech, Szostak showed that the mere presence of RNA in fatty-acid vesicles could promote their growth by appropriating membrane molecules from neighboring vesicles that contain less RNA or none at all.[27] This growth occurs because the RNA contained within vesicles exerts an osmotic pressure on these sacs. This internal pressure places tension on the membrane, which grows by absorbing fatty acids from any surrounding vesicles that are

less swollen as a consequence of having less genetic material. The protocells with more RNA inside them would grow faster, to the point
where, with just a little shaking—which is consistent with the action of
wind or waves on primordial Earth—they would break into daughter
vesicles.

Szostak's next step was once again to insert RNA but this time to
endow this software with useful instructions for his protocells. That
information could encode the means to make phospholipids, a class of
lipids that feature in modern membranes. This would have been a critical step in the transition from primitive membranes, which are based
on fatty acids, to modern cell membranes, which are based on phospholipids. By introducing RNA software into protocells this way, simple, self-replicating systems are theoretically possible. These are
exciting studies, and I am sure they will demonstrate the feasibility of
self-replicating cells forming out of prebiotic chemicals.

If a synthetic genetic material can be designed to catalyze its own
reproduction within an artificial membrane, life of a primitive type
will have been created in the laboratory. Perhaps these cells will resemble the first forms of life on Earth, from nearly four billion years
ago, but more likely they will represent something quite new. Impor-
tantly, these early synthetic cells, much like those at the dawn of life,
would be brimming with potential: they would be subject to mutation
and Darwinian evolution. I am sure that when it comes to the ambitious goal of turning DNA into a cell, they will provide valuable insights that will dovetail with the efforts of my own team, and those of
many others exploring these profound issues.

Complementing this work on the origins of life, we are conducting
new research that has the long-term aim of creating a "universal recipient cell" that can take any synthetic DNA software customized to create life and create that designated species. At present the number of
types of recipient cells that we use for genome transplantation in our
laboratories is very limited. In order to create a universal recipient cell,
we are in the process of rewriting the genetic code of the mycoplasma
cell to enable it to transcribe and translate any transplanted DNA

software. This research should refine and extend our insights into why life comes in the little packages we call cells.

In a more radical approach, we are investigating how to do away with the need for an existing cell as a recipient for a synthetic genome. Our hope is that we can create synthetic cells, starting with cell-free systems and then adding basic components to construct a complete cell. Although this could be groundbreaking, once again, related research dates back a surprisingly long time. As early as the beginnings of the DNA revolution, in the 1950s, several research teams had independently demonstrated that a cell is not strictly necessary when it comes to carrying out some of the basic transactions of life. They found that the manufacture of proteins can take place even after the membrane of a cell has disintegrated.

The possibilities were first proposed by Paul Charles Zamecnik, a professor of medicine at Harvard Medical School and a senior scientist at nearby Massachusetts General Hospital. He became interested in the subject in 1938, when, at the autopsy of a morbidly obese woman, he was struck by the abundant amount of fat in her tissue but the relative paucity of protein. It caused him to wonder how proteins were manufactured, a question that would drive his research for much of his career.[28] He would come to realize that, in order to work out the intermediary events of protein synthesis, he would need to develop a cell-free system. After several years of effort, he finally achieved this with the aid of his colleague, Nancy Bucher, paving the way to many significant insights, from revealing that ATP was required for protein synthesis to discovering that ribosomes were the site of protein assembly.

Many research groups have worked on reconstructing biological processes from isolated components. In 1955 Heinz Fraenkel-Conrat and Robley C. Williams were the first to show, using tobacco mosaic virus, that a functional virus could be created out of purified RNA and a protein coat. The deciphering of the basics of the genetic code, and how information gets translated from DNA software to protein, quickly followed, as a result of the pioneering work of Marshall Nirenberg and his postdoctoral fellow, J. Heinrich Matthaei, in 1961.[29] In their experiment they prepared an extract from bacterial cells that could make protein even when no intact living cells were present. Using synthetic RNAs and radioactively labeled amino acids, they discovered that three uracil's, UUU, formed the codon for the amino acid phenylalanine.

Since then it has been a matter of routine to take DNA or RNA and produce proteins in vitro. As a result, cell-free protein synthesis has become an important tool for molecular biologists. Although these methods traditionally required cell extracts, this changed with the introduction of what is called the PURE system (protein synthesis using recombinant elements), in which protein synthesis is done in a cell-free system,[30] with the translation machinery of *E. coli* reconstituted from purified components and ribosomes. We are attempting to use a cocktail of enzymes, ribosomes, and chemicals (including lipids) with a synthetic genome to create new cells and life forms without the need for pre-existing cells. In years to come, it will be increasingly possible to create a wide variety of cells from computer-designed software of life in cell-free systems and or via a universal recipient cell.

The eventual creation of cells from scratch will open up extraordinary new possibilities. First, as we explore this boundary between the animate and inanimate, we should be able to refine our definition of "life." This work could also have consequences for how we define such words as "machine" and "organism."[31] The ability to create life without pre-existing cells will also have very practical implications, since we will be able to increase the degree of freedom in the design of new forms of life. We could explore older varieties of life as well, by deducing the genome of an extinct creature from the genomes of its living descendants and using synthetic cells to explore the properties of this ancient software.

We will also begin to explore the potential of assemblies of synthetic cells. The human body is itself a remarkable collective endeavor, with the digestive system alone being home to roughly one hundred trillion microbes—about ten times the number of cells in all the major organs of your body. The vast majority are friendly microbes that work cooperatively with our biochemistry. This propensity for cells to work together began relatively early in the history of life. Multicellular strings of bacteria arose around 3.5 billion years ago. There are other forms of microbial cooperation, as mentioned earlier. The late Lynn Margulis, at the University of Massachusetts, Amherst, proposed that specialized eukaryotic cells acquired their photosynthesis or energy-producing mitochondria by symbiogenesis, the helpful merger of two ancestral cells.

Such early examples were followed by another wave of cooperation, when these complex cells themselves teamed up to form communities, and evolved to do so independently several times. More than six hundred million years ago, ctenophores—common and fragile jellies with well-developed tissues—marked the point at which multicellular life began to diversify. Sponges are another early example of solitary cells that came to cooperate in more complex bodies. They consist of various different kinds of cells—digestive cells, cells that secrete the spicules (segments of the body skeleton), and so on—which can communicate with one another and work together as a single individual.

The genome sequence of one sponge, *Amphimedon queenslandica*, a Great Barrier Reef demosponge, reveals some of the genetic mechanisms that allowed individual cells to work together.[32] There are half a dozen hallmarks of multicellularity: regulated cell cycling and growth; programmed cell death (apoptosis); cell–cell and cell–matrix adhesion, which enables the component cells to cling together; developmental signaling and gene regulation; mechanisms to defend against invading pathogens; and the specialization of cell types, which is why we have nerve cells, muscle cells, and so on. Given how many times multicellularity evolved independently, it seems unlikely that there is a single explanation for its origins, save that cellular cooperation has been the best solution to the evolutionary question of how to be more successful at passing genes to the next generation, whether that meant defending against particular parasites or developing a more efficient way to move about and exploit available sources of food and energy.

With the rise of the synthetic cell, we can lay bare details of the mechanisms that led to multicellularity. Synthetic cells could be stripped down and simplified to see how each of the multicellular factors listed above can influence the ability of cells to communicate and cooperate. This would give us an unprecedented tool to untangle the enormously complicated interactions that occur between cells in a multicellular creature, whether a nematode worm or a human being. At the same time, there will be attempts to build synthetic multicellular creatures from the bottom up, with synthetic cells containing synthetic organelles, so we can investigate this intimate form of cooperation.

As early as the late 1960s a team at the State University of New York, Buffalo, had successfully created a relatively large organism, *Amoeba proteus,* from the major cellular components from other amoebae: nucleus, cytoplasm, and cell membrane.[33] They reported how the "success of our reassembly experiments means that we now have the technical ability to assemble amoebae which contain any desired combination of components and thus have an excellent test system." We can give these manufactured cells more efficient cellular batteries, or create a synthetic endoplasmic reticulum, the organelle into which ribosomes are studded and where protein synthesis and folding take place.

From our studies of the mycoplasmas and other organisms, we have already identified the ingredients for the basic recipe of a living cell: a cocktail of around three hundred to five hundred or so proteins (roughly the same number as emerged from Lucy Shapiro's work on the "essential genome" of the bacterium *Caulobacter crescentus*). Imagine if we can systematically explore variants on the machinery of life, learning which components are crucial, which are not, and teasing apart how they work together. This will be a boon for the field of synthetic biology by expanding the range of biological constituents, software subroutines, and circuits that we can develop.

10 Life by Design

*A new variety raised by man will be a far more important and interest-
ing subject for study than one more species added to the infinitude of
already recorded species.*

—Charles Darwin, *On the Origin of Species* (1859)[1]

When we design and write new software to program living cells, how
can we increase our confidence that it will work? The obvious way is to
actually attempt to create the cell, but currently that is relatively expensive and time-consuming, and if you fail you'll be left wondering
whether the problem lay in the software itself or in the boot-up system
that turns the DNA's instructions into reality. In the future, computational modeling will offer one way to put our understanding to the test
by creating virtual cells before attempting real ones. Computer modeling of life is coming closer to having an impact than ever before, in
part because of the exponential rise of the power of computers and in
part because modern biology has seen a proliferation of studies that
generate enormous troves of information. Over the past two decades,
in particular, the scientific community has gathered increasing
amounts of detailed data on biological systems, from systems analysis
to the intricately folded three-dimensional structures of proteins.
There is a wonderful variety of molecular machines, with a wide range
of functions, and we are learning more about how they interact with
one another and with other cellular components. As a result of this
data deluge, a wide range of basic biology can now be modeled *in silico*
to complement laboratory experiments.

Efforts by a number of teams to computationally simulate life have
been under way for decades at various degrees of sophistication,

including modeling biochemical processes such as gene regulation to simulations of metabolism and protein synthesis. In Europe, for example, the Virtual Physiological Human project[2] aims to model the workings of organs in a computer to create a virtual body. To do this successfully, they have had to integrate a broad range of knowledge about physiology, from the tens of thousands of genes and their variants to the much greater numbers of protein components to changes in metabolism.

Attempts to model organs and tissue have been tried for some time. The first mathematical models of cardiac cells appeared in 1960.[3] By the 1980s the activities—electrical, chemical, and mechanical—involved in the contraction of a heart-muscle cell were reasonably well understood, and it became possible to create a computer model of a beating heart cell. Around thirty equations captured the key cellular chemical processes, notably the action of the ion channels that allow electrical signals to flicker in and out of cardiac cells. The increase in computer power since then has made it possible to simulate the beating of the billions of these cells in all four chambers of a virtual heart.[4]

The quest to simulate organs has also targeted the brain, with its billions of interconnected neurons. The Human Brain Project,[5] at the École Polytechnique Fédérale de Lausanne, on the shores of Lake Geneva, had by 2008 simulated a microcircuit, which consists of a unit of ten thousand brain cells in the cerebral cortex, the thin layer in the brain where the most interesting and advanced functions of thinking reside. To simulate a human brain with its one hundred billion neurons will take at least a decade or more, and at the start of 2013, the European Commission announced that it would spend one billion euros on the project.[6]

At a more fundamental level, a number of efforts have been made over the years to create an *in silico* virtual cell—a dynamic biological system in the form of "living" software, one that might show how all the processes inside a living cell function together as one system. Although I have not been directly involved in these projects, they have benefitted from my institute's work. The profound knowledge and insights that we had gained in our genome research on mycoplasmas made it possible for others to create a detailed model of a *Mycoplasma* cell in a computer.

In the 1990s, an attempt to turn our genomic data into an "electronic cell" was made by a team led by Masaru Tomita, at Keio University, in Fujisawa, Japan. When the Keio group began its project, only eighteen organisms had been sequenced. They believed that the then-unprecedented amount of molecular information available for a wide range of model organisms would yield vivid new insights into intracellular molecular processes that could, if simulated in a computer, enable them to predict the dynamic behavior of living cells. Within a computer it would be possible to explore the functions of proteins, protein–protein interactions, protein–DNA interactions, regulation of gene expression, and other features of cellular metabolism. In other words, a virtual cell could provide a new perspective on both the software and hardware of life.

In the spring of 1996 Tomita and his students at the Laboratory for Bioinformatics at Keio started investigating the molecular biology of *Mycoplasma genitalium* (which we had sequenced in 1995) and by the end of that year had established the E-Cell Project. The Japanese team had constructed a model of a hypothetical cell with only 127 genes, which were sufficient for transcription, translation, and energy production. Most of the genes that they used were taken from *Mycoplasma genitalium*. In their simulation the team charted out a spider's web of metabolic interactions of this hypothetical genome, which included twenty tRNA genes and two rRNA genes. The cell relied on what they admitted were unrealistically favorable environmental conditions.

The state of the modeled cell at any given moment was expressed as a list of concentrations of its component substances, along with values for cell volume, acidity (pH), and temperature. To model DNA software the team used real software routines and developed hundreds of rules that governed many but not all of the metabolic pathways of *M. genitalium*, including glycolysis, lactate fermentation, glucose uptake, glycerol and fatty-acid uptake, phospholipid biosynthesis, gene transcription, protein synthesis, polymerase and ribosome assembly, along with protein degradation and mRNA degradation. For extra fidelity, the model was constructed so that enzymes and other proteins degraded spontaneously over time, so that they had to be constantly synthesized in order for the cell to sustain "life."

The Japanese team conducted experiments on the virtual cell as the "simulator engine" chugged away at around one twentieth the rate of a real living organism. They could "starve" the virtual cell by draining glucose from the culture medium. When this was done they observed how the quantity of ATP temporarily rose but then fell sharply until, after running out of ATP fuel, the cell would eventually "die." If glucose was then added back, the virtual cell might or might not recover, depending on the duration of its starvation. The model could, at the click of a mouse, replicate the effect of knocking out a gene on the concentration levels of a range of cellular substances, millisecond by millisecond. The cell could also be "killed" by knocking out an essential gene controlling, for example, protein synthesis. As a consequence, all enzymes would gradually degrade and eventually disappear altogether.

But at that time, in the late 1990s, it still presented a challenge to link the various levels of cellular processes, from the use of genes to the metabolism and beyond. The Keio model's 127-gene format was also much smaller than the "minimal gene set" derived from our gene knockouts and through sequence comparison of our first two sequenced genomes. As a result, the model cell was "self-supporting," but it was not capable of proliferating; it lacked pathways for DNA replication, gene regulation, and the cell cycle. And, of course, at that time, as today, there were/are still many genes whose functions were not yet known, so the scientists had to rely on informed guesswork to fill in missing metabolic functions.

The Japanese team has come a long way in the decade since their programming of the original E-Cell. As well as constantly refining the model, they have moved on to modeling human red blood cells (erythrocytes), neurons, and other kinds of cells, and to other challenges that complement this work on virtual cells, such as measuring the way that *E. coli* responds to genetic and environmental challenges, revealing that the web of metabolic interactions within the cell is surprisingly robust, as a result of redundancies.[7]

The most recent work on *Mycoplasma genitalium* has been done in America, by the systems biologist Markus W. Covert, at Stanford University. His team used our genome data to create a virtual version that came remarkably close to its real-life counterpart. This tour de force depended on the synthesis of a vast amount of information, including data from more than nine hundred scientific papers, on the organism's genome, transcriptome, proteome, metabolome, and any other "-ome" you can think of. As a result, *M. genitalium* became the first organism to be modeled in detail, down to every one of its 525 genes and every known gene function.

To create their virtual cell, the Stanford team used several thousand parameters related to around thirty modules of subcellular processes, each of which was modeled in various ways. To integrate them into a unified cellular machine, they programmed the separate modules, each governed by its own algorithm, to communicate with one another. In this way the simulated bacterium was created as a series of modules that mimic the various functions of the cell. Using a network of 128 computers, the Stanford team could chart the behavior of virtual *M. genitalium* cells at the molecular level, from their DNA and RNA to the proteins and metabolites.[8] They modeled the life span of the cell at the molecular level, charting the interactions of twenty-eight categories of molecules. Finally, the overall model was validated against a knowledge base that consisted of databases of information about *M. genitalium*.[9]

The Stanford team also used their virtual organism to investigate details of the cell cycle, which consists of three stages—initiation, replication, and cytokinesis (cell division). They noticed larger variability in the first stages, compared with the last, and when compared with the duration of the overall cell cycle. The length of individual stages in the cell cycle varied from virtual cell to cell, while the length of the overall cycle was much more consistent. Consulting the model, they hypothesized that the overall cell cycle's lack of consistency was the result of a built-in negative-feedback mechanism, which compensated for differences in the individual stages. Cells that took longer to begin DNA replication had time to amass a large pool of free nucleotides. The actual step of replication, in which those nucleotides were used to create new DNA strands, then passed relatively quickly. Cells that went through the initial step more quickly, on the other hand, had no nucleotide surplus. Replication then ended up being delayed by the rate of nucleotide production.

The Stanford team ran simulations of the effects of mutations for all 525 genes to determine if the mutated cell remained viable. Their predictions were around 80 percent accurate, when compared with experimental data on physical cells. When there were differences in the results, however, interesting insights emerged. According to the model, deletion of the gene lpdA should have killed the cell; in fact, this strain does remain viable, though it grows 40 percent more slowly than the wild type. The team reasoned that another protein must have been performing a task similar to that of lpdA. On closer inspection, they found that the relevant gene, nox, was similar to lpdA in terms of sequence and function. When they corrected their model of a virtual cell by adding in this additional function of nox, it yielded a viable simulated cell as well. Discrepancies in growth rates of mutants observed *in silico* and in real life helped the team to fine-tune the rate at which enzymes were produced in its silicon cells, once again to make them more realistic and more like *M. genitalium*.

Such models will ultimately give us the freedom to undertake "what if" scenarios, a common approach in the field of engineering. Just as an engineer adjusts the width of a structural component of a skyscraper on a computer to see what happens to its ability to withstand an earthquake, systems biologists can manipulate the software of life to explore its effect on the viability of cells. It will be interesting to compare our computer-designed minimal genome to the computer models to see how predictive gene changes can be.

Realizing this revolution in biological computing is going to require a vast surge in computer power. At the moment, running a simulation for a single cell to divide once takes the Stanford team around ten hours and generates half a gigabyte of data. The first virtual bacterium, with its 525 genes, is far less complex than *E. coli*, which has 4,288 genes, divides every twenty to thirty minutes, and has a much greater number of molecular interactions, each of which would further expand the time required to run a simulation. More complex eukaryotic cells will present significant challenges in creating virtual versions.

There is still much work to be learned from real organisms to understand how the linear software code maps out the three-dimensional world of the cell. One important effort in that direction has been led by another Stanford scientist, Lucy Shapiro, who undertook an unusual career move, from fine-arts major to biological researcher. She has since spent her career in developmental biology, focusing on an asymmetrically organized freshwater bacterium called *Caulobacter crescentus*. In 2001 my team collaborated with Lucy's to determine the genetic code of *Caulobacter*, which has 4,016,942 base pairs coding for 3,767 genes.[10]

Lucy Shapiro's work has demonstrated that bacteria are not just disorganized bags of proteins but have distinct intracellular compartments, with specific protein robots occupying specific sites to orchestrate complex biochemical processes, such as the cell cycle and division. Shapiro revealed for the first time that bacterial-DNA replication occurs in a spatially organized way and that cell division depends on this organization and on segregation of the DNA to opposite ends of the cell. Her team also proved the existence of master genetic regulators of the cell cycle. For example, a regulatory gene involved in building the flagellum, the whip-like appendage that enables the organism to swim, is also essential to viability. They found that different events, which they had been studying as if they were isolated in the bacterial life cycle, were actually connected by global regulators. Just one of these regulators controlled the expression of ninety-five other genes. In a streamlined version of the work we attempted with *Mycoplasma*, Shapiro has catalogued, down to the letter, exactly what parts of the genetic code are essential for *Caulobacter crescentus* to survive—around 12 percent of the bacterium's genetic material. The essential elements include not only protein-coding genes but also regulatory DNA and, intriguingly, ninety-one small DNA segments of unknown function. The other 88 percent of the genome could be disrupted without harming the bacterium's ability to grow and reproduce.[11]

The future of biological research will be based to a great extent on the combination of computer science and synthetic biology. We can get a fascinating view of this future from a series of contests that culminate in a remarkable event that takes place each year in Cambridge, Massachusetts—a gathering of brilliant young minds that gives me great hope for the future. The International Genetically Engineered Machine (iGEM) competition invites high school and college students and entrepreneurs to shuffle a standard set of DNA subroutines into something new in a competition for a trophy, a large aluminium Lego brick that symbolizes the belief that life can built by snapping together subroutines.

The competition was first devised by three engineers who wanted to apply the Lego-like logic of building systems from interlocking parts to biology: Tom Knight, Randy Rettberg, and Drew Endy. Now held at MIT, the event grew out of a course offered there in January of 2003, in which teams were challenged to design a form of *E. coli* that "blinked"—that is, generated fluorescent light at regular intervals. That evolved into a summer competition attended by five teams in 2004 and thirteen teams in 2005—the first year that the competition went international. Since then it has grown rapidly, with 245 teams in 2012.

For iGEM 2011, around 160 teams and over two thousand participants from thirty countries worldwide took part in the competition, which began with regional heats and ended with a world championship. The event was not a dry college course with PowerPoints, lectures, and demonstrations but a genetics jamboree in which contestants have mascots and wear team T-shirts with sponsorship logos as they slice and dice DNA software.

One aim of this endeavor is to build up a catalogue of standardized parts: namely, connectable pieces of DNA called BioBricks, which program a host bacterium to perform a specific task. Each BioBrick is capped at both ends with DNA sequences that enable it to be connected to other bricks and integrated into a plasmid that can be inserted into a bacterial cell. Over the years the participants have amassed a centralized, open-source genetic library of thousands of BioBricks, called the Registry of Standard Biological Parts. The registry contains a list of function, structure, and so on, and is itself modeled on a thousand-page catalogue of circuit components called *The TTL Data Book for Design Engineers*.

For each competition, the teams are given a particular set of biological parts—a BioBrick kit—at the beginning of the summer. Copies of the actual DNA are mailed to them in the form of dehydrated DNA. Working at their own schools or labs over the summer, they use these parts and new parts of their own design to construct biological systems and operate them in living cells. In the quest to assemble new biological "circuits" from a set of standardized parts, some teams of students have attempted the kinds of approaches that my own team has been investigating.

Basic items in their toolkit of life include promoters, which mark which segments of DNA are to be read; operators, which can regulate the use of promoters; ribosome binding sites, which recruit ribosomes to begin making proteins; the protein coding sequence itself, which could code for an enzyme, a repressor to bind to and disable a promoter, or a reporter, such as a green fluorescent protein, which can, as the name suggests, reveal that a circuit is active; and terminators, which give the signal to stop reading the DNA software. These parts can be assembled into devices that, in a cell, carry out simple functions.

At its most basic, a device can make a protein. But because DNA is the software of life, it can also be used to build logic gates, the elementary building blocks of computers—such as an AND gate (where both of two inputs must be present for a gene to be turned on); an OR gate (where either one or another input is needed to turn on a gene); or a NOT gate, where a protein is not made when a signal is present, and vice versa. Devices can also send signals between cells to coordinate the behavior of cell populations, as bacteria naturally do with so-called quorum sensing, which enables them to tune gene activity according to the number of bacterial neighbors. There are also light-control devices, which rely on light-gathering proteins, such as photoreceptors from plants and bacteria.[12] The circuits constructed from these gene gates may one day be components of engineered cells that can monitor and respond to their environments.

In turn, devices can be linked together in a system. For example, they can produce feedback loops, either to generate positive feedback (using an activator), of the kind that turns a low sound in a microphone into a howl, or negative feedback (using a repressor), of the kind that is used in a thermostat to turn off a heater when a given temperature is reached. One can build switches,[13] which respond to the conditions within a cell or around it, using a promoter and repressor, or oscillators, which follow a cyclical pattern (think of your body clock) that can be built in various ways, such as combining a negative-feedback loop with a delay, or counters, where an event triggers the production of a protein, which in turn activates another protein generator.

In this way, the synthetic-biology student can construct a hierarchy, starting with parts and moving to devices and then systems. As a result of their work, we now have cellular circuits capable of pattern generation, noise shaping, edge detection,[14] event counting, and synchronized oscillations in a growing population of cells.[15] One team from Cornell has designed a cell-free method for producing complex biomolecules called, naturally enough, a BioFactory. A bacterium can be made to fluoresce in a clockwork fashion by creating a loop of genetic instructions. *E. coli* can be converted into an information storage device, a "bio-hard" disk, as was done by a team from the Chinese University of Hong Kong, who called their work *E. cryptor*. Others have devised software to manipulate DNA software on screen and then turn it into sequences in the lab, with the help of robots.[16]

More playful projects include bacteria that glow in the dark and, in the case of the MIT "Eau d'e coli" project, that smell like wintergreen while they're growing and like bananas when they stop. There are living LCDs—computer screens made of yeast or bacterial cells instead of digital pixels. One team, from the University of Texas, Austin, and University of California, San Francisco, displayed the phrase "Hello World" with *E. coli*[17] engineered to sense light, using a cyanobacteria protein domain to control the gene lacZ, which can cleave a molecule to produce black pigment.

Teams have also created all kinds of altruistic microbes, including those designed to change color in response to an environmental toxin, do computing,[18] reveal parasites,[19] detect land mines, or, in the case of common yeast, yield beta carotene, the orange substance that gives carrots their color. One team from Cambridge University, in the UK, devised *E. chromi* to paint *E. coli* in vivid colors, a feat that has won nominations for art and design awards.[20] Others have brewed "BioBeer," rich in resveratrol, a chemical found in wine that is thought by some to have health benefits.

The competition is very aware of the societal aspects of synthetic biology and the need to have non-scientists understand and accept their attempts to tinker with the machinery of life. The competitors are deeply involved, as part of their projects, in interacting with the public, taking surveys, and speaking with the press. Safety is also central, and each team has to write a report on the impact of their project on the world around them. One contestant went as far as developing algorithms to determine how similar a specific DNA sequence was to entries in the Centers for Disease Control and Prevention's Select Agent and Toxin List.[21]

I applaud the BioBrick approach for drawing in and teaching students, and have nothing but praise for the iGEM originators. I think they have changed the course of education in this field. The extraordinary ingenuity demonstrated in iGEM projects gives me hope for the future by encouraging a new generation of scientists to experience the excitement of manipulating the software of life. We have moved a long way from blindly altering genomes by selective breeding, as in traditional agriculture and farming, toward the design of life with modern science.

While these students are learning biological design, impressive advances are being made regularly by many talented researchers at many different laboratories around the world. Some are developing laboratories on chips—biochips—to combine in situ protein synthesis, assembly, and imaging to create patterns of proteins.[22] Others have learned how to make proteins from a single copy of DNA within a microscopic quartz chamber.[23] As shown so vividly by iGEM, there is also a global effort to recapitulate genetic circuits in vitro.[24] For the future of genome design we will need toolkits of new artificial amino acids, on-and-off switches, biological rheostats, oscillators, modulators, suicide genes, and gene pathways for engineering life.

Let me focus on one or two examples, to give a glimpse of the potential. To read the commands in DNA software, cells rely on zinc finger proteins, commonly known as "zinc fingers." They were discovered in 1985 by Aaron Klug, a Nobel laureate working at the Medical Research Council's Laboratory of Molecular Biology, in Cambridge.[25] These proteins are so named because they contain a zinc atom and are shaped like an index finger.[26] There are hundreds of varieties, all of which work by binding to DNA where each finger matches up with a three-letter sequence of DNA. The more fingers you use, the more precisely you can recognize a particular sequence. With just six fingers, you can target any particular gene.

This important piece of biological machinery has been adapted for synthetic biology by Boston University biomedical engineers Ahmad S. Khalil and James J. Collins. They have created novel zinc-finger designs that are intended to bind with new target sequences.[27] The Boston team has engineered new circuits in yeast, a eukaryote, using modular, functional parts from the eukaryotes themselves and "wiring" them up with the help of zinc fingers.[28] There are many immediate applications of this work, such as helping to develop stem cells for regenerative medicine, and in-cell devices and circuits for diagnosing early stages of cancer and other diseases. This method may also equip groups of cells to perform higher-order computational tasks for processing signals in the environment in sensing applications.

Efforts by others are under way to extend and modify the existing genetic code, to code for new amino acids that don't exist in nature. The genetic code is redundant, as in some cases multiple codons code for the same amino acid. Those extra codons can be allocated to encode new amino acids that are not present in nature among the twenty "standard" amino acids. In one such study, Jason Chin at the Medical Research Council's Laboratory of Molecular Biology, engineered fruit flies that incorporated three new amino acids into proteins in the cells of their ovaries. The use of new amino acids includes the addition of new functions to proteins and making cells resistant to viral infections.

But most important of all, a systematic exploration of the potential of synthetic biology will deepen our understanding of fundamental biology. With such capabilities, we can expand our knowledge of biology thousands of times faster than is possible today. The insights we achieve will in turn help improve genome designs that can be tested in virtual cell models and thus will accelerate the effort to synthesize new life.

Designing and creating new life forms continues to raise a number of important ethical questions. These have been explored in numerous initiatives, not just in America but in many other countries with mature biotech industries. I initiated the first ethical review of synthetic genomes and synthetic life in the late 1990s, when my institute funded the University of Pennsylvania's bioethics department to review our work. My team has worked with various government agencies, including the U.S. Department of Energy, the White House Office of Science and Technology Policy (OSTP), and the National Institutes of Health since our work on phi X 174. In 2004, for just one example, our policy team, headed by Robert Friedman, along with the Center for Strategic and International Studies (CSIS) and MIT, was funded by the Alfred P. Sloan Foundation to conduct workshops and a public session over a twenty-month period to discuss the ethical and societal implications of synthetic genomics. We (I was among the core members, which included George Church, Drew Endy, Tom Knight, and Ham Smith) published our findings in October of 2007 as *Synthetic Genomics: Options for Governance.*[29]

All the while I give public lectures, as do members of my team; present detailed papers to academic conferences; and deal with constant questions from the global media. We have made several trips to Capitol Hill to brief more than fifty members of Congress and have addressed the OSTP, the CIA, the National Science Advisory Board for Biosecurity (NSABB), the Presidential Commission for the Study of Bioethical Issues, and the Department of Homeland Security. Reports on synthetic biology have been issued by many bodies, such as the U.S. Department of Energy and the NSABB. Public consultations have been sponsored, as well, not only in America but in countries such as the United Kingdom.[30] Delegates from leading institutions and associations came together at the OECD/U.S. National Academies/UK Royal Society Symposium, in July 2009, and considered the opportunities, threats, and wider questions posed by synthetic biology, such as what it means to be human. From any perspective, the discussions concerning precisely what it means to create synthetic life have been long, full, and open.

One point that has struck me forcefully over the years is that few of the questions raised by synthetic genomics are truly new. One of the more famous attempts to grapple with issues raised by synthetic life forms can be found in the Three Laws of Robotics, which were devised by the science-fiction author Isaac Asimov and first appeared in 1942 in his short story "Runaround": "1. A robot may not injure a human being or, through inaction, allow a human being to come to harm. 2. A robot must obey the orders given to it by human beings, except where such orders would conflict with the First Law. 3. A robot must protect its own existence as long as such protection does not conflict with the First or Second Laws." Asimov would later add a fourth, "the zeroth law," to precede the others: "0. A robot may not harm humanity, or, by inaction, allow humanity to come to harm." One can apply these principles equally to our efforts to alter the basic machinery of life by substituting "synthetic life form" for "robot."

Emerging technologies, whether in robotics or in synthetic biology, can be a double-edged sword. Today there is much debate about "dual-use" technologies—as, for instance, in a study published in 2012 by the American Association for the Advancement of Science, a group of American universities, and the FBI.[31] Their investigation was prompted by research undertaken by teams in America and the Netherlands that had identified the elements of the H5N1 influenza virus that enabled its rapid spread. When each team's results were submitted to *Science* and *Nature*, in August 2011 they prompted widespread unease. Such concerns led scientists worldwide to voluntarily agree, in a self-imposed moratorium, to stop these studies to understand and curb pandemic threats until they understood how to proceed with and communicate the research safely.

The problem here, of course, is that while such studies could help identify viruses that pose the greatest threat to human life, and develop treatments for them, they also provide information that could be abused by terrorists. The U.S. National Science Advisory Board for Biosecurity reviewed the dual-use implications and recommended that both H5N1 papers be published only after researchers had removed key data. But a February 2012 meeting by the World Health Organization concluded that the benefits of the work outweighed the risks and expressed doubts about redacting the papers. Later an FBI report offered a number of suggestions to get the balance right between making progress with research and minimizing risks, and between scientific freedom and national security.

The FBI report begins by pointing out that the Janus-like nature of innovation has surfaced again and again during the past several decades, underscoring the significance of such initiatives as Asilomar, which I dealt with earlier, and the adoption of the Biological and Toxin Weapons Convention of 1972. I believe that the issue of the responsible use of science is fundamental and dates back to the birth of human ingenuity, when humankind first discovered how to make fire on demand. (Do I use it to burn a rival's crops or to keep warm?) Every few months, another meeting is held to discuss the conundrum that powerful technology cuts both ways.

But it is important not to lose sight of the opportunities that this research presents. Synthetic biology can help address key challenges facing the planet and its population, such as food security, sustainable energy, and health. Over time, research in synthetic biology may lead to new products that will produce clean energy and help quell pollution; help us grow crops on more marginal land; and provide more affordable agricultural products, as well as vaccines and other medicines. Some have even speculated about the ability of smart proteins or programmed cells to self-assemble at the sites of disease to repair damage.

Clearly, this apparently limitless potential raises many unsettling questions, not least because synthetic biology frees the design of life from the shackles of evolution and opens up new vistas for life. It is crucial that we invest in underpinning technologies, science, education, and policy in order to ensure the safe and efficient development of synthetic biology. Opportunities for public debate and discussion on this topic must be sponsored, and the lay public must engage with the relevant issues. I hope that, in some small way, this book will help readers to make sense of the spectrum of recent developments.

Safety, of course, is paramount. The good news is that, thanks to a debate that dates back to Asilomar in the 1970s, robust and diverse regulations for the safe use of biotechnology and recombinant-DNA technology are already firmly in place. However, we must be vigilant and never drop our guard. In years to come it might be difficult to identify agents of concern if they look like nothing we have encountered before. The political, societal, and scientific backdrop is continually evolving and has shifted a great deal since the days of Asilomar. Synthetic biology also relies on the skills of scientists who have little experience in biology, such as mathematicians and electrical engineers. As shown by the efforts of the budding synthetic biologists at iGEM, the field is no longer the province of highly skilled senior scientists only. The democratization of knowledge and the rise of "open-source biology"; the establishment of a biological design-build facility, BIOFAB,[32] in California; and the availability of kitchen-sink versions of key laboratory tools, such as the DNA-copying method PCR,[33] make it easier for anyone—including those outside the usual networks of government, commercial, and university laboratories and the culture of responsible training and biosecurity—to play with the software of life.

There are also "biohackers" who want to experiment freely with the software of life. The theoretical physicist and mathematician Freeman Dyson[34] has already speculated on what would happen if the tools of genetic modification became widely accessible in the form of domesticated biotechnology: "There will be do-it-yourself kits for gardeners who will use genetic engineering to breed new varieties of roses and orchids. Also kits for lovers of pigeons and parrots and lizards and snakes to breed new varieties of pets. Breeders of dogs and cats will have their kits too."

Many have focused on the risks of this technology's falling into the "wrong hands." The events of September 11, 2001, the anthrax attacks that followed, and the H1N1 and H7N9 influenza pandemic threat have all underscored the need to take their concerns seriously.[35] Bioterrorism is becoming ever more likely as the technology matures and becomes ever more available. However, it is not easy to synthesize a virus, let alone one that is virulent or infective,[36] or to create it in a form that can be used in a practical way as a weapon. And, of course, as demonstrated by the remarkable speed with which we can now sequence a pathogen, the same technology makes it easier to counteract with new vaccines.

For me, a concern is "bioerror": the fallout that could occur as the result of DNA manipulation by a non-scientifically trained biohacker or "biopunk."[37] As the technology becomes more widespread and the risks increase, our notions of harm are changing, along with our view of what we mean by the "natural environment" as human activities alter the climate and, in turn, change our world.

In a similar vein, creatures that are not "normal" tend to be seen as monsters, as the product of an abuse of power and responsibility, as most vividly illustrated by the story of Frankenstein.[38] Still, it is important to maintain our sense of perspective and of balance. Despite the knee-jerk demands for ever more onerous regulation and control measures consistent with the "precautionary principle"—whatever we mean by that much-abused term[39]—we must not lose sight of the extraordinary power of this technology to bring about positive benefits for the world.

I am not alone in believing that overregulation can be as harmful as laxness in that regard. I was glad to see that my own view was echoed in the response to my work on the first working synthetic genome, when the Presidential Commission for the Study of Bioethical Issues released a report in December 2010[40] entitled *New Directions: The Ethics of Synthetic Biology and Emerging Technologies*. This document opened with a letter from President Barack Obama that emphasized how vital it was that, as a society, we consider in a thoughtful manner the significance of this work, and strike a balance between "important benefits" and "genuine concerns."

The commission was led by political theorist Amy Gutmann, president of the University of Pennsylvania, and included experts in bioethics, law, philosophy, and science. In its conclusions the commission identified five guiding ethical principles that it considered relevant to the social implications of emerging technologies: public beneficence, responsible stewardship, intellectual freedom and responsibility, democratic deliberation, and justice and fairness. If those principles were diligently used to illuminate and guide public-policy choices as we advanced with synthetic biology, the commission concluded, we could be confident that the technology could be developed in a responsible and ethical manner.

Among its recommendations to the president, the commission said that the government should undertake a coordinated evaluation of public funding for synthetic-biology research, including studies on techniques for risk assessment and risk reduction and on ethical and social issues, so as to reveal noticeable gaps, if one considered that "public good" should be the main aim. The recommendations were, fortunately, pragmatic: given the embryonic state of the field, innovation should be encouraged, and, rather than creating a traditional system of bureaucracy and red tape, the patchwork quilt of regulation and guidance of the field by existing bodies should be coordinated.

Concerns were, of course, expressed about "low-probability, potentially high-impact events," such as the creation of a doomsday virus. These rare but catastrophic possibilities should not be ignored, given that we are still reeling from the horrors of September 11. Nor should they be overstated: though one can gain access to "dangerous" viral DNA sequences, obtaining them is a long way from growing them successfully in a laboratory. Still, the report stated that safeguards should be instituted for monitoring, containment, and control of synthetic organisms—for instance, by the incorporation of "suicide genes," molecular "brakes," "kill switches," or "seatbelts" that restrain growth rates or require special diets, such as novel amino acids, to limit their ability to thrive outside the laboratory. As was the case with our "branded" bacterium, we need to find new ways to label and tag synthetic organisms.

More broadly, the report called for international dialogue about this emerging technology, as well as adequate training to remind all those engaged in this work of their responsibilities and obligations, not least to biosafety and stewardship of biodiversity, ecosystems, and food supplies. Though it encouraged the government to back a culture of self-regulation, it also urged it to be vigilant about the possibilities of do-it-yourself synthetic biology being carried out in what it called "noninstitutional settings." One problem facing anyone who casts a critical eye over synthetic biology is that the field is evolving so quickly. For that reason, assessments of the technology should be under rolling review, and we should be ready to introduce new safety and control measures as necessary.

In acknowledgement of how society needs to embrace the vision of synthetic biologists if democracy is to truly function, the report also called for scientific, religious, and civil engagement, public education, and the exchange of views about both the promise and the perils—without recourse by bloggers and journalists to lazy sensationalism (well-worn criticisms about "playing God"), partial reporting, or distortions. I would be the first to agree that we have to work hard and listen carefully to the public and remain vigilant in order to earn their trust.

There will always be Luddites who believe we should not go down this path at all, who would rather we gave up the effort to create synthetic life and turned our backs on this "disruptive technology." In 1964 Isaac Asimov made a wise remark about the rise of robots that could apply equally to the rise of redesigned life: "Knowledge has its dangers, yes, but is the response to be a retreat from knowledge? Or is knowledge to be used as itself a barrier to the dangers it brings? With all this in mind I began, in 1940, to write robot stories of my own—but robot stories of a new variety. Never, never, was one of my robots to turn stupidly on his creator for no purpose but to demonstrate, for one more weary time, the crime and punishment of Faust."[41] My greatest fear is not the abuse of technology but that we will not use it at all, and forgo a remarkable opportunity at a time when we are overpopulating our planet and changing environments forever. If we abandon a technology, we abandon the means to use it to improve and save lives. The consequences of inaction can be more dangerous than the improper use of technology.

I can envisage that, in the coming decades, we will witness many extraordinary developments of tangible value, such as crops that are resistant to drought, that can tolerate disease and thrive in barren environments, that provide rich new sources of protein and other nutrients, that can be harnessed for water purification in harsh and arid regions. I can imagine designing simple animal forms that provide novel sources of nutrients and pharmaceuticals, customizing human stem cells to regenerate a damaged, old, or sick body. There will be new ways to enhance the human body as well, such as boosting intelligence, adapting it to new environments such as the radiation levels encountered in space, rejuvenating worn-out muscles, and so on.

Let's keep our focus on the global problems that are affecting humanity. Many serious issues now threaten our fragile and overcrowded world, one that will soon be home to nine billion people, one that is running out of fundamental resources such as food, water, and energy, and one that is haunted by the specter of unpredictable and devastating climate change.

11 Biological Teleportation

There was a sharp click and the man had disappeared. I looked with amazement at Challenger. "Good heavens! Did you touch the machine, Professor?"

—Arthur Conan Doyle, "The Disintegration Machine" (1929)[1]

Many of the greatest and most revolutionary ideas, from moon rockets to invisibility, have been anticipated by myth, legend, and, of course, science fiction. This is also true of our efforts to use our understanding of the software of life to transport the digital instructions to build a living organism, or one of its components, from one place to another on this planet, or even between planets or to locations far beyond our own solar system.

The enduring idea of the transporter, which disassembles people or objects at one location and reassembles them elsewhere, was popularized by Gene Roddenberry (1921–1991) in his 1960s television series *Star Trek*. ("Scotty, beam us up.")[2] The transporter was born out of a prosaic problem facing Roddenberry: he lacked the budget to show a starship landing during each week's episode. That same decade, British TV audiences were introduced to Doctor Who and his TARDIS (time and relative dimension in space), a blue London police box that could transport its occupants to any point in time and any place in the universe.

The idea of teleportation did not originate with *Star Trek* or *Doctor Who* but has been in one form or other a part of literature for centuries. In *One Thousand and One Nights* (often known as *The Arabian Nights*), a collection of stories and folk tales compiled during the Islamic Golden Age and published in English in 1706, genies (djinns) can transport themselves and objects from place to place instantaneously. Arthur Conan Doyle's "The Disintegration Machine," published in 1929, describes a machine that can atomize and reform objects. Teleportation has been explored by many fantasy and science-fiction writers, including Isaac Asimov ("It's Such a Beautiful Day"),[3] George Langelaan ("The Fly"), J. K. Rowling (the Harry Potter series), and Steven Gould (*Jumper*). While these presentations of teleportation are purely fictional, the concept of "quantum teleportation" is very much a reality and was introduced to a wider audience by Michael Crichton in his 1999 novel *Timeline*, which was later turned into a movie.

The origins of quantum teleportation date back much earlier and rest, in part, on an intellectual disagreement between two of Schrödinger's most impressive peers in the development of the theory of the atomic world (quantum theory): Albert Einstein, who disliked the theory's strange take on reality, and Niels Bohr (1885–1962), the Danish father of atomic physics. In 1935, in the course of this dispute, Einstein highlighted one perplexing feature of quantum theory with the help of a thought experiment that he devised with his colleagues Boris Podolsky (1896–1966) and Nathan Rosen (1909–1995).

They first noted that quantum theory applied not only to single atoms but also to molecules made of groups of atoms. So, for example, a molecule containing two atoms could be described by a single mathematical expression called a wave function. Einstein realized that if you separated these constituent atoms by a vast distance, even placing them at opposite ends of the cosmos, they would still be described by the same wave function. In the jargon, they were "entangled." More than half a century later, in 1993,[4] Charles H. Bennett, of IBM, and others theorized that pairs of entangled atoms in effect established a "quantum phone line" that could "teleport" all the details (quantum state) of one particle to another an arbitrary distance away, without knowing its state. This opened up the possibility that a transporter could transmit atomic data. Experiments have followed to establish that this is indeed possible. The record for long-distance quantum teleportation at the time of this writing is by an international research team using the European

Space Agency's Optical Ground Station, in the Canary Islands, which reproduced the characteristics of a light particle across 143 kilometers of open air.[5] The experiment saw the teleportation of the states of light particles, or photons, between La Palma and Tenerife.

Teleportation also has the potential to enable a new kind of computer—a quantum computer—to operate and solve problems millions of times faster than current computers.[6] A group from Caltech in 1998 reported the first experimental demonstration of the teleportation of the quantum state of a light beam.[7] The feat was at first demonstrated between single photons, between a photon and matter, and between single ions (charged atoms). Then in 2012 the first teleportation of macroscopic objects—large enough to see—was reported, between two atomic ensembles, each consisting of about one hundred million rubidium atoms and measuring around one millimeter across, linked by a 150-meter optical fiber. The team that reported the feat—led by Jian-Wei Pan, of the Hefei National Laboratory for Physical Sciences at the Microscale, at the University of Science and Technology of China, in Hefei—said this technique could be used to transfer and exchange information in future quantum computers and networks, leading to speculation about a "quantum internet."[8]

However impressive these advances, the reality of *Star Trek* teleportation remains a far-off prospect. In an interview with *Scientific American*, one of the pioneers who did the first experiment in 1998, H. Jeff Kimble, of Caltech, was asked to describe the biggest misconception about teleportation: "That the object itself is being sent. We're not sending around material stuff. If I wanted to send you a Boeing 757, I could send you all the parts, or I could send you a blueprint showing all the parts, and it's much easier to send a blueprint. Teleportation is a protocol about how to send a quantum state—a wave function—from one place to another."[9] You would need something on the order of 10^{32} bits of information about his atoms to successfully teleport a human being.

But, of course, as Kimble suggests, you *can* transmit digitized instructions or software. A human genome contains only around 6×10^9

bits of information. My team is perfecting a way to send the digitized version of DNA code in the form of an electromagnetic wave—and then use a unique receiver at a distant location to re-create life. This would mark a transformation between two fundamental domains of particle types. All the life that we know of on Earth is a chemical-based system whose every structural component—DNA, RNA, proteins, lipids, and other molecules—is composed of individual atoms of different chemical elements (carbon, hydrogen, oxygen, iron, and so on). The elements and their own building blocks (for example, the electrons, which orbit the nucleus, and quarks, which make up the nucleus) are collectively called fermions. ("Fermions," a reference to the great Enrico Fermi (1901–1954), is a term coined by the English physicist Paul Dirac (1902–1984), who shared the Nobel Prize in Physics for 1933 with Erwin Schrödinger "for the discovery of new productive forms of atomic theory.") The other generic class consists of bosons, which includes the Higgs and all particles that carry forces, notably gluons, the W and the Z, and the photon, the stuff of electromagnetic waves. The key difference between fermions and bosons is a quantum property called "spin." Bosons have, by definition, integer spin; quarks, electrons, and other fermions all have a half unit of spin. This results in a huge difference in their behavior and, in the case of fermions, is responsible for the whole of chemistry, and therefore biology.

When we read the genetic code by sequencing a genome, we are converting the physical code of DNA into a digital code that can be transformed into an electromagnetic wave that can be transmitted at the speed of light. It was Dimitar Sasselov, director of the Harvard Origins of Life Initiative, who drew my attention to how this feat straddles the two great domains of particles:

> So, life as we know it, and as it seems to have originated historically on our planet, is a fermionic phenomenon—all its structures are made of fermions. The information coded in the DNA molecule is coded with the help of fermions, and is read with the help of fermions. Our ability today to represent that information in digital form and

transmit it over distances using electromagnetic waves (at the speed of light!) marks life's transition from purely fermionic to bosonic.[10]

At Synthetic Genomics, Inc. (SGI), we can feed the digital DNA code into a software program that automatically works out how to re-synthesize the sequence in the laboratory. This automates the process of designing overlapping oligonucleotides of fifty to eighty base pairs, adding unique restriction sites and watermarks, and then feeds them into the integrated oligonucleotide synthesizer. The synthesizer will rapidly produce the oligonucleotides, which will be automatically pooled and assembled using our Gibson-assembly robot.

Although oligonucleotide synthesis can be carried out with remarkably higher fidelity than was possible forty years ago, it remains an error-prone process that does produce a fraction of unintended DNA sequences, a fraction that grows with the size of the synthetic DNA fragments. The synthesis error rate for the assembly of standard oligonucleotides is generally about one error per thousand base pairs. Thus, it should be expected from this rate that if oligonucleotide errors are not weeded out early on in the building process—by, for example, cloning and sequencing, or with an error-correction enzyme—most, if not all, DNA fragments above ten thousand bases will contain errors. To deal with this fundamental problem, we have come up with a new approach that should pave the way to high-fidelity DNA synthesis.

Following oligo assembly and PCR amplification, we can now remove any DNA containing errors with an enzyme called endonuclease. This particular biological robot was discovered by using a software system called Archetype, which was developed by Toby Richardson and his team at SGI for storing, managing, and analyzing biological sequence data. The "error correction" process begins by denaturing and annealing the DNA amplified by PCR, so that it forms double-stranded DNA. A few of the double-stranded DNA molecules contain the correct DNA sequence at every position and are ignored by the endonuclease. However, in DNA where there has been a substitution, deletion, or insertion, double-stranded DNA with base-pair mismatches forms, known as heteroduplex DNA. This is recognized and cleaved by the endonuclease.

The fact that intact molecules amplify more efficiently than endonuclease-digested DNA means that we can use a second PCR reaction to enrich the percentage of error-free synthetic fragments. This approach generally yields much lower error rates, better than one per fifteen thousand synthesized base pairs, and can be further improved by performing additional rounds of error correction. At this stage we have produced a DNA molecule with sufficient precision that it can be the final product in its own right, such as a DNA vaccine (where DNA is introduced into cells in the body to make a vaccine protein). But the potential is limitless. With synthesized DNA, it will eventually be possible to create all forms of life.

Using in vitro cell-free protein synthesis of the kind pioneered by Marshall Nirenberg in the 1960s, synthetic DNA constructs can now be utilized to produce proteins in an automated system. The DNA from a phage or virus needs only be introduced into a receptive bacterial cell, where it will take over that cell's protein and DNA synthesis machinery and make many copies of itself.

Sometimes it can be hard to see beyond the horizon of present potential at the moment that a technology such as the "biological teleporter" is crystallizing from an idea into something real. That was certainly the case when it came to the laser, which was initially billed as a solution looking for a problem.[11] But I think we can already perceive how our future will be shaped by the ability to translate the software of life into light. The ability to send DNA code to anywhere on the planet in less than a second holds all kinds of possibilities when it comes to treating disease and illness. This information could code for a new vaccine, a protein drug (such as insulin or growth hormone), a phage to fight an infection caused by a resistant strain of bacterium, or a new cell to produce therapeutics, food, fuel, or clean water. When combined with home synthesizers, this technology will also allow treatments to be customized for each and every person, so that they suit the genetic makeup of a patient and, as a result, minimize side effects.

The most obvious immediate application is to distribute vaccines in the event of the appearance of an influenza pandemic. The last such outbreak was announced on June 11, 2009, when the World Health Organization declared H1N1 influenza (swine flu) to be the first pandemic in more than forty years, triggering an international response to address this major public-health threat. The result was the fastest global vaccine-development effort in history. Within six months, hundreds of millions of vaccine doses had been produced and distributed around the world, demonstrating the ability for rapid mobilization and cooperation between public- and private-sector institutions worldwide.

But despite this unprecedentedly rapid response, it was not fast enough. Meaningful quantities of the vaccine were available only two months after the peak of viral infection, leaving the majority of the population exposed to the pathogen at the height of its circulation. Even though the death rate was relatively low, vast numbers of people were exposed to the virus. Around 250,000 people died from H1N1 and, due to the nature of this particular influenza, most of them were young. Had this virus been more pathogenic, the lag time in vaccine availability might have resulted in a severe health crisis, one that might well have led to strife, disorder, and social breakdown in affected cities.

A century ago just such a severely pathogenic flu strain did sweep around the planet, with devastating consequences. Worldwide, the death toll from the 1918–1920 pandemic—about fifty million people—was greater than that of the First World War. One doctor said it was "the most vicious type of pneumonia that has ever been seen." Using mortality data from that pandemic, a team led by Christopher Murray, of Harvard University, predicted in the *Lancet* that sixty-two million people—96 percent from the developing world—could die within a year if a similar pandemic were to occur today. The recent swine flu pandemic was a call to arms and has highlighted the necessity of getting vaccines to people quickly.

Synthetic Genomics, Inc., and the J. Craig Venter Institute have announced a three-year collaboration agreement with Novartis to apply synthetic-genomics tools and technologies to accelerate the production of influenza seed strains. The seed strain is the starter culture of a virus, a live reference virus, and is the base from which larger quantities of the vaccine virus can be grown. The agreement, supported by an award from the U.S. Biomedical Advanced Research and Development Authority (BARDA), could ultimately lead to a more effective response to both seasonal and pandemic flu outbreaks.

Currently Novartis and other vaccine companies rely on the World Health Organization to identify and distribute the seed viruses. To speed up the process we are using a method called "reverse vaccinology," which was first applied to the development of a meningococcal vaccine by Rino Rappuoli, now at Novartis. The basic idea is that the entire pathogenic genome of an influenza virus can be screened using bioinformatic approaches to identify and analyze its genes. Next, particular genes are selected for attributes that would make good vaccine targets, such as outer-membrane proteins. Those proteins then undergo normal testing for immune responses.

My team has sequenced genes representing the diversity of influenza viruses that have been encountered since 2005. We have sequenced the complete genomes of a large collection of human influenza isolates, as well as a select number of avian and other non-human influenza strains relevant to the evolution of viruses with pandemic potential, and made the information publicly available. The strains have been chosen to represent many subtypes with a wide geographical and chronological distribution. As a result of our collaboration, Novartis and SGI will have developed a "bank" of synthetically constructed seed viruses that are ready to go into production as soon as the World Health Organization identifies the circulating flu strains. This technology could reduce the vaccine production time by up to two months, which would be of critical benefit in the event of a pandemic.

Standard influenza vaccine manufacture is a time-consuming process. An important rate-limiting step is the lag between strain selection—when the WHO and Centers for Disease Control, in Atlanta, identify the circulating strain(s) and make a global recommendation for the creation of the specific influenza seed virus—and the actual production of the vaccine. The traditional method of manufacture relies on growing the viruses in fertilized hen eggs. In all it takes around thirty-five days for the process, which includes testing and distribution of a reference virus, co-infection in eggs with standard backbone viruses, and isolation and purification of the vaccine seeds. By taking advantage of major advancements in synthetic biology and cell-based manufacturing, and by introducing the exciting concept of digital-to-biological conversion, we and Novartis have produced better quality vaccines in fewer than five days.

The vaccine is based on the viral envelope proteins hemagglutinin (HA), which forms spikes that enable the influenza virus to attach to target cells, and neuraminidase (NA), which forms knoblike structures on the surface of virus particles and catalyzes their release from infected cells, allowing the virus to spread. Once high-fidelity synthetic HA and NA genes have been produced, the next step is to "rescue" the complete vaccine seed by combining the HA and NA with the other handful of genes in the influenza genome, which has the software to make just eleven proteins. We used a reverse-genetics approach with a Novartis MDCK (Madin-Darby canine kidney) cell line, approved by the U.S. Food and Drug Administration in 2012 to replace eggs for flu-vaccine manufacture. The cells are infected with linear synthetic cassettes encoding the relevant genes. Influenza viruses can be detected in cell-culture mediums within seventy-two hours following transfection, and the desired viral strain can be isolated for further amplification and eventual use as vaccine seeds.

A proof-of-concept test was performed on August 29, 2011, to demonstrate the proficiency and robustness of the synthetic vaccine-seed production process. Starting with HA and NA gene sequences from a low-pathogenicity North American avian H7N9 influenza strain provided by BARDA and obtained from the Centers for Disease Control, oligonucleotide synthesis was initiated at 8 A.M. By noon on September 4, exactly four days and four hours from the start of the procedure, the seed virus was produced. Since the completion of this initial proof-of-concept demonstration, the process has been successfully repeated for multiple additional influenza strains and subtypes, including H1N1, H5N1, and H3N2. At the time of this writing, no strains have been encountered that cannot be assembled and rescued synthetically. In 2013, our DNA-assembly robot made the H7N9 genes without any human intervention.

The synthetic influenza project is now entering the important next phases of development. Rapid and efficient vaccine-seed production reveals new opportunities for pandemic preparedness when combined with an understanding of how the virus evolves. Influenza viruses are dynamic and change in two basic ways: antigenic drift and antigenic shift. The drift, which is constant, refers to small, gradual changes that occur through point mutations in the two genes that contain the genetic material to produce the main surface proteins mentioned above, hemagglutinin and neuraminidase. Antigenic shift refers to an abrupt, major change to produce a novel influenza virus, either through direct transmission to humans from animals (such as pigs or birds) or through mixing of human influenza A and animal influenza A virus genes to create a new human influenza A subtype virus, through a process called genetic reassortment. By monitoring and analyzing influenza strain shifts and novel emerging strains, we can begin to prepare for predicted strains, rather than scrambling to react when an actual outbreak occurs. For strains that represent potential threats, vaccine seeds can be made in advance and kept in viral-seed banks, available for off-the-shelf use when needed.

Preparations for the next pandemic are well under way. A comprehensive database that will include past influenza sequence data, antigenic variation, and strain growth characteristics is currently being built and is already populated with almost sixty-six thousand isolate sequences and thirty-three thousand pairwise sets of antigenic data. Advanced algorithms are being developed to predict changes in the proportion of circulating virus subpopulations over time, to generate estimates for the best vaccine (and candidate vaccine) strains to provide protection, and to improve the predictive capability of strain selection.

Another critical step toward advanced synthetic-based influenza vaccines is the integration and scale-up of synthetic vaccine seeds with full-scale manufacturing to enable commercial vaccine production. Novartis is starting to make this a reality—a reality that has enormous implications for the global pandemic response. The speed, ease, and accuracy with which higher-yielding influenza vaccine seeds can be produced using synthetic techniques promises not only more rapid future pandemic responses but a more reliable supply of pandemic influenza vaccines.

While vaccines are the best means of prevention against pandemics, and synthetic biology has helped us to make them more effective, we are now facing another major threat from infection, as one of humankind's most important weapons in the fight against disease, antibiotics, are rapidly becoming compromised. We have enjoyed a ceasefire in the historic war with microbes since the middle of the last century, after penicillin was accidentally discovered in 1928 by British microbiologist Alexander Fleming (1881–1955), and a method to mass-produce the drug was developed by the Australian Howard Walter Florey (1898–1968), along with the German Ernst Boris Chain (1906–1979) and the English biochemist Norman Heatley (1911–2004).[12] Along with Fleming, Florey and Chain shared the 1945 Nobel Prize in Medicine for their far-reaching work. During the past eight decades, antibiotics have been used to cure a range of once-deadly infectious diseases, saving millions of lives, and have greatly extended the use of surgery—imagine attempting the removal of an appendix without them, let alone a heart, kidney, or hip.

Although this family of drugs has had a colossal effect on extending life span, from the outset the microbes they target have been fighting back. Soon after penicillin was used to treat soldiers in the Second World War, bacteria had already evolved ways to resist this common antibiotic. Insights into how bacteria could shrug off a given antibiotic came from elegant experiments carried out at the University of Wisconsin by Joshua Lederberg (1925–2008), who was inspired to study bacterial genetics by Avery, MacLeod, and McCarty's seminal 1944 paper that identified DNA as the "transforming principle." Working alongside his wife, Esther Zimmer Lederberg, he showed that strains of bacteria resistant to penicillin were already present naturally before penicillin was used as a drug, part of a range of important research that would win him the Nobel Prize.

Resistant strains use a wide range of proteins to neutralize the effects of antibiotics. Antibiotics such as tetracycline and streptomycin bind to a specific region of a ribosome to disrupt protein synthesis, and one way microbes have evolved to thwart these drugs is to manufacture ribosomes that do not bind to the drugs. Some microbes have evolved "efflux pumps," proteins that can eject an antibiotic before it has time to act. Some resistant strains cloak themselves in impermeable membranes. Other microbes even "eat" antibiotics.[13] There are so many mechanisms of resistance[14] that some even refer to the "resistome."[15]

Because bacteria divide so quickly, any resistant strains soon come to dominate a population. They also use another mechanism to spread resistance: they can swap their DNA software in a process called "lateral gene transfer," also called horizontal gene transfer. Lederberg showed one way they could do this, via cell-to-cell contact or a bridge-like connection.[16] At the molecular level they swap plasmids that could contain several antibiotic-resistance genes. If this transfer is successful, a superbug is born.

The development of resistant organisms is inevitable but, unfortunately, has been spurred by poor infection control, much of which boils down to hygiene and hand-washing, or lack thereof. The rise of resistance has also been driven by the indiscriminate use of antibiotic drugs, notably on the farm; misuse, in treating viral infections such as the common cold; underuse, when a course of the drug is not completed; and overuse, in soaps and other domestic products. As if that weren't bad enough, current market conditions provide little incentive for companies to undertake the arduous effort of developing novel antibiotics. Unlike heart drugs and other medicines, antibiotic drugs are only taken for a week or so. The relentless rise of resistance means antibiotic drugs are all destined to become useless after a while, so the shelf life of a new antibiotic is limited.

This is an eloquent example of Darwinian evolution, albeit one with a depressing message: the golden age of antibiotics may be at an end. There are countless examples of the advance of resistance: one persistent stalker of hospital wards, methicillin-resistant *Staphylococcus aureus*, has acquired full resistance to vancomycin, which is often billed as a last-ditch treatment. For the past few years the fear has been expressed again and again that we may be facing a return to a pre-antibiotic era,[17] when the greatest cause of death was disease caused by bacteria and the local hospital was a hotbed of infection, the last place that you would want to be if you really wanted to get better.

Genomics can help a great deal. We can chart the rise of a superbug, learn how it defies antibiotics, and find new targets for drugs. We can also draw on synthetic genomics to produce alternatives to antibiotics. One approach that we are pursuing is to revisit an antibacterial treatment called phage therapy, in which bacteriophages that are specific to a certain bacterial strain are used to kill the microbe. Every few days, half the bacteria on Earth are killed by phages.[18] Can we enlist their help in the fight against superbugs?

The bacteriophages, which are ten times more abundant than bacteria, were discovered—perhaps by two individuals independently[19]—around a century ago. The first to identify them, in 1915, was English scientist Frederick Twort (1877–1950), an eccentric polymath who made violins, radios, and more, and who also tried to breed the biggest sweet pea in England.[20] The French-Canadian microbiologist Félix d'Herelle (1873–1949) enters the phage story in 1917, and first used the term bacteriophages ("eaters of bacteria") to describe them. He maintained that the phenomenon described by Twort was something quite different.[21] Speculating that bacteriophages played a role in recovery from dysentery, d'Herelle recognized their potential in fighting infection[22] and conducted the first human study in 1919. After the phage preparation was ingested by d'Herelle and colleagues at high doses in order to confirm its safety, he administered a diluted phage preparation to a twelve-year-old boy with severe dysentery who recovered within a few days.

D'Herelle's studies helped to explain a puzzling observation: what was in the waters, such as those found in the sewage-ridden Ganges and Yamuna rivers, in India, that provided protection from cholera?[23] Now the answer was clear. A drop of river water or sewage teems with millions of phages. By the 1930s phage cocktails were being manufactured by companies in Europe and America for the treatment of many infections. Two of the most prominent laboratories were d'Herelle's, in France, and another that he co-founded in Tbilisi, in the Soviet republic of Georgia, in 1923. The lab was named the Eliava Institute of Bacteriophage, Microbiology and Virology after its co-founder, the Georgian phage researcher George Eliava (1892–1937), who had the blessing of Soviet dictator Joseph Stalin. Due in part to his collaboration with foreign scientists, including d'Herelle, and for pursuing a woman who was also admired by Lavrentiy Beriya, Stalin's chief of secret police, Eliava was pronounced an "enemy of the people" and executed in 1937.[24] The Eliava Institute survived without its founder and became one of the largest units developing therapeutic phage, at its peak producing several tons a day. In 1989 Mikhail Gorbachev, the last president of the Soviet Union, restored Eliava's name during a reassessment of the victims of the Great Purge.

By the middle of the 1930s, the hype and hope that phage therapy would end bacterial diseases had failed to materialize, and any evidence of its efficacy had been clouded by the lack of standardization of materials.[25] During that decade the American Medical Association issued withering critiques of the method,[26] but lying as they did at the borderline of life, phages continued to fascinate basic researchers. As one sign of the importance attached to this work, Alfred Hershey and Salvador Luria even set up the "Phage Church" with Max Delbrück to focus on the fundamentals of biological replication, and thus of heredity.

Phage therapy was used by the Soviet and German armies in the Second World War, but with the rise of antibiotics and the end of that conflict, in the West it, along with all things Communist, would come to be regarded as suspect in the postwar era.[27] One disciple of Delbrück's "church," Gunther Stent, wrote in 1963 how "the strange bacteriophage therapy chapter of the history of medicine may now be fairly considered as closed. Just why bacteriophages, so virulent in their antibacterial action in vitro, proved so impotent in vivo has never been adequately explained."[28]

One reason is that the history of phage therapy—much like the history of pretty much anything else you can think of—is "rich with politics, personal feuds, and unrecognized conflicts."[29] But a more significant one is that it had to wait for the rise of modern scientific methods to refine the therapy. Fortunately, by the time the Soviet Union was dissolved, in 1991, the Eliava Institute was still supplying phages for newly independent Georgia, and significant work was being done at the Ludwik Hirszfeld Institute of Immunology and Experimental Therapy, in Wroclaw, Poland. Today, as the arms race with microbes continues to shift in their favor, a great many researchers—including my team—are reevaluating the use of phages to fight infections.[30]

Unlike traditional antibiotics, which can cause collateral damage by killing vast numbers of "friendly" bacteria in our bodies, such as those that enable us to digest our food, phages are like molecular "smart bombs," targeting only one or a few strains or substrains of bacteria. We now have a detailed picture of how these microbial killing machines can attack a single kind of bacterium with surgical precision. Take for example the T4 phage, which has been studied by many pioneers in the field, from Max Delbrück and Salvador Luria to James Watson and Francis Crick. The one hundred sixty-nine thousand bases of its genome contain all the instructions necessary to infect and destroy the microbe *E. coli.*

Measuring around ninety nanometers wide and two hundred nanometers long, the T4 phage is large compared with other phages and looks like a microscopic space lander, with "legs" that attach to specific receptors on the surface of an *E. coli* cell, and a hollow tail that can inject its software into the bacterium. (It has only recently been discovered that T4 can pierce the cell membrane with an iron-tipped spike.[31]) Although the injected DNA is quite different from that of its host, the coding language is the same, so the target bacterium carries out the

instructions to build a phage, killing itself in the process. After manufacturing some one hundred to one hundred fifty phages, the bacterium bursts, releasing a horde of newly made phages into the surrounding
environment.

As is the case with antibiotics, cells can mutate to survive and develop resistance to phages.[32] Humans also clear phages rapidly from
the bloodstream. But phages do appear to offer an interesting alterna-
tive to antibiotics. To date, those that have been used in therapy have
been isolated from the environment, including from sewage, and science has been limited to the range of natural phages. However, with
our new DNA synthesis and assembly tools, we could design and synthesize hundreds of new phages per day, or synthesize over five thousand new variations on a sequence theme. With this unique capacity
we will be able to test and realize the dream of d'Herelle.

These technologies will enable a complete and rapid bacteriophage
design cycle that stretches from isolation to characterization, engineering, and evolution, leading to the assembly of libraries of opti-
mized therapeutic phages for clinical use to fight superbugs. As we
demonstrated with phi X 174, selection by infectivity—where a target
bacterium obligingly multiplies phages best suited to infect it—is a
very powerful tool that will easily enable high-throughput screening of
newly synthesized phages for the desired traits of broad or highly focused spectrums of effectiveness.

Perhaps in the future we will be able to sequence an infectious agent
from an individual patient to identify the target microbe and quickly
design a custom phage therapy. The new phage could be sent instantly
to the patient or a therapeutic center or hospital with the teleporting
technology I have described. Synthetic phages could also be engineered
to be maximally effective. For example, they could be designed to target proteins and gene circuits in superbugs other than those affected
by conventional drugs for use alone or in combination with antibiotics.
We believe that we can make more potent versions of a powerful antibacterial killing machine called lysin, which helps a phage break out of
an infected cell. One lysin called PlyC (streptococcal C1 phage lysin)

kills bacteria more quickly than bleach and consists of nine protein parts that assemble to form what looks like a flying saucer, which locks on to the bacterial surface using eight separate docking sites located on one side of the saucer.[33] The two "warheads" of PlyC chew through the wall of the cell, killing the bacterium and releasing the phage. Lysins have been developed that control a wide range of Gram-positive pathogens such as *S. aureus, S. pneumoniae, E. faecalis, E. faecium, B. anthracis*, and group *B. streptococci.*

Due to the great specificity of phages, one would also expect them to be safe. In August 2006 the FDA approved spraying meat with a phage preparation targeted against *Listeria monocytogenes*, created by the company Intralytix. The following year, an early clinical trial was completed at the Royal National Throat, Nose and Ear Hospital, London, for *Pseudomonas aeruginosa* infections of the ear (otitis), and the results were promising.[34]

The potential value of using new synthetic phages to treat drug-resistant infections is likely to be realized because of the accelerating pace of developments in the field of synthetic genomics and technology to transmit genetic information almost immediately. But we will still need the rigor of modern methods to pull phage therapy out of its pseudoscientific past. Such therapies may be controversial because they are cocktails of viruses that have the potential to multiply and evolve. They do, after all, have a role in disease by arming bacteria with genes, such as one linked with diphtheria, so safety approval will require some careful work. Still, I suspect that these concerns will rapidly abate when this approach begins to show promise, such as in veterinary medicine, or for treating common disorders such as acne.[35]

To help realize the potential to curb infectious disease, my team is already testing methods of transmitting and receiving DNA software. NASA has funded us to carry out experiments at its test site in the Mojave Desert, which straddles California, Nevada, Utah, and Arizona. We will be using the JCVI mobile laboratory, which is equipped with soil-sampling, DNA-isolation, and DNA-sequencing equipment, to document and test all of the required steps for autonomously isolating microbes from soil, sequencing their DNA, and then transmitting the information to the cloud with what we call a "digitized-life-sending unit."

I have no doubt that this technology will work. We have been doing a more primitive version of remote sampling around the globe for the past decade, mostly on the *Sorcerer II* expedition, named after the yacht I used to voyage around the world's oceans. We have logged more than eighty thousand miles at sea, taking samples every two hundred miles, and if we had the kind of technology described above, it would have been possible to sequence as we sailed. Instead, due to the current fragility of laboratory sequencers, we had to rely on Federal Express and UPS to deliver the samples to our laboratories.

To complement our efforts to build a digitized-life-sending unit, we are building a receiving unit, where the transmitted DNA can be reproduced anew. This device has a number of names at present, including "digital biological converter," "biological teleporter," and—the preference of former *Wired* editor in chief Chris Anderson—"life replicator." Creating life at the speed of light is part of a new industrial revolution that will see manufacturing shift away from the centralized factories of the past to a distributed, domestic manufacturing future, thanks to 3-D printers. This technology is already being used to assemble embryonic stem cells into tissues, grow bones, and to build planes or even entire buildings by "concrete printing." Why stock warehouses with parts when entire designs can now be stored in virtual computer warehouses waiting to be printed locally and on demand? We might one day get to a point where individuals can make all the products they want, from door handles to smartphones, including the next generation of 3-D printer. Soon, you may be able to take a picture of a broken part from a washing machine, television, or any other appliance with your smartphone, and then pay for the license to print a replacement at home. As a result, the cornerstones of consumer culture—the shopping mall and the factory—will become increasingly irrelevant.

The key economic considerations in this scenario will become the raw materials and the intellectual-property costs. As for the benefits, I think the most revolutionary could be the use of specialized printers for biological manufacturing. At this point in time we are limited to making protein molecules, viruses, phages, and single microbial cells, but the field will move extremely quickly to more complex living systems. There are already home versions of 3-D printers, and various groups are already looking at the use of modified ink-jet printers to print cells and organs. This is a fascinating area that works by layering living cells on a structural matrix in the shape of a blood vessel or a human organ. Whatever we end up calling these devices, I am confident that in coming years we will be able to convert digitized information into living cells that will become complex multicellular organisms or can be "printed" to form three-dimensional functioning tissues. The ability to print an organism remains some way off but will become a possibility soon enough. We are moving toward a borderless world in which electrons and electromagnetic waves will carry digitized information here, there, and everywhere. Borne upon those waves of information, life will move at the speed of light.

12 Life at the Speed of Light

*The changing of Bodies into Light, and Light into Bodies, is very con-
formable to the Course of Nature, which seems delighted with Transmu-
tations.*

—Sir Isaac Newton, *Opticks* (1718)[1]

When life is finally able to travel at the speed of light, the universe
will shrink, and our own powers will expand. Simple calculations indicate that we could send electromagnetic sequence information to a
digital-biological converter on Mars in as little as 4.3 minutes, at the
closest approach of the Red Planet, to provide a settlement of colo-
nists with vaccines, antibiotics, or personalized drugs. Likewise,
if, for example, NASA's Mars *Curiosity* rover was equipped with a
DNA-sequencing device, it could transmit the digital code of a Martian microbe back to Earth, where we could re-create the organism in
the laboratory.

This latter approach to searching for extraterrestrial life rests on
two important assumptions. First, that Martian life is, like life on
Earth, based on DNA. I think that is a reasonable supposition, because
we know that life existed on Earth close to four billion years ago and
that the Earth and Mars have continually exchanged material. Planets
and their satellites in the inner solar system—including Earth—have
been sharing materials for billions of years, as rocks and soil from successive collisions with asteroids and comets have been thrown into
space.[2] Chemical analysis confirms that meteorites found on Earth
must have been blasted off the surface of the Red Planet by an asteroid
impact. Simulations suggest that only 4 percent of material ejected
from Mars reaches our planet, after a journey that can take as long as

fifteen million years. Even so, it has been estimated that Earth and Mars exchange on the order of a hundred kilograms of material a year, making it likely that every shovel of terrestrial earth contains traces of Martian soil. It is therefore likely that Earth microbes traveled to and populated Martian oceans long ago and that Martian microbes survived to thrive on Earth.

Second, and more fundamental, is my assumption that life does indeed exist elsewhere in the universe. There are still many people (often religious) who believe that life on Earth is somehow special, or unique, and that we are alone in the cosmos. I am not among them.

Scientists have a great deal of confidence that Mars will prove to contain life or have contained life. So much so that they, and the media, have been inclined to be somewhat trigger-happy when interpreting evidence from the Red Planet. In Chapter 3, I recounted the furor that followed the 1996 publication of a paper that detailed the evidence that had convinced some NASA scientists of the existence of Martian microbial life. The supposed life traces observed in the meteorite in question, known as ALH 84001,[3] were far from being the first ambiguous signals of this kind. In 1989 a team led by Colin Pillinger, at the Open University, Milton Keynes, UK, found organic material, typical of that left by the remains of living things, in another Mars meteorite, EETA 79001,[4] though they stopped short of announcing they had discovered life on Mars. Others have discerned hazy indications of life after reevaluating the data collected by NASA's *Viking* landers, which conducted the first in situ measurements focusing on the detection of organic compounds when they touched down on the Red Planet in 1976.[5]

At the end of 2012 there was much feverish speculation about what had been found by the Sample Analysis at Mars, or SAM, instrument on the *Curiosity* rover when it studied soil grains from a wind drift named Rocknest. Weeks earlier a scientist on the project had inadvertently set off expectations of a momentous revelation when he told National Public Radio that the data was "one for the history books."

The disappointment was palpable when, that December, the American Geophysical Union conference, in San Francisco, was told by the instrument scientists that there was indeed evidence of organic compounds but that more work had to be done to determine whether they were indigenous to Mars.[6] While these data do reveal possible hints of Martian life, we need extraordinary evidence if we are going to make any extraordinary claims. I am confident that life once thrived on Mars and may well still exist there today in subsurface environments. Compelling data suggest that liquid water flowed on the surface of the planet, including possible oceans,[7] and clays around Matijevic Hill indicate the water there might once have been pure enough to drink.[8] In late 2012 *Curiosity* found signs of an ancient streambed where water once flowed rapidly.[9] Today, however, it appears to exist in a frozen state, including at the polar ice caps and in the form of permafrost. There is increasing evidence for substantial subsurface water on Mars that is frozen and, though speculative, that liquid water exists deeper in the planet.[10] Calculations estimate that brine water could be found at a depth of four kilometers and pure liquid water at a depth of eight kilometers.[11] Subsurface Mars also contains substantial methane,[12] which might also be of biological origin, though we can't rule out that its source could be geological, or a combination of the two.

I have already been involved in the quest to identify subterranean life. One of our teams at Synthetic Genomics, in collaboration with BP, spent three years studying life in coal-bed methane wells in Colorado. We found remarkable evidence, in water samples from one mile (1.6 kilometers) deep, of the same density of microbes as are found in the oceans (one million cells per milliliter). However, the subterranean organisms were much less varied in terms of species diversity, most likely due to the lack of oxygen (all the deep subterranean cells are anaerobic) and ultraviolet radiation, the key producers of genetic mutations. One of the fascinating consequences of these conditions, including the low rate of evolution, was our discovery that the genome sequences of one of these organisms closely matched those of a microbe isolated from a volcano in Italy. While subterranean species diversity is still high, when we look at any one class of organisms there is only 1 to 3 percent variation, while in the oceans we see variation of up to 50 percent in, for example, SAR11, the most abundant marine photosynthetic microbe.

What we discovered in the depths of the planet was a range of extremophiles that are able to use the carbon dioxide and hydrogen present underground to produce methane in a manner similar to that of the *Methanococcus jannaschii* cells isolated near a "smoker," a hydrothermal vent 2,600 meters deep in the Pacific Ocean. Simple calculations indicate that there is as much biology and biomass in the subsurface of our Earth as in the entire visible world on the planet's surface. Subterranean species have likely thrived there for billions of years.

If one accepts that liquid water is synonymous with life, Mars should be inhabited by similar organisms. Evidence is increasing that Mars had oceans three billion years ago and again perhaps as recently as one billion years ago, when the polar ice caps melted following a meteor impact. Evidence from the many Mars rovers and probes suggests that, though habitable environments once existed on the planet, they probably dried up several billion years ago.

Radiation levels are much higher on Mars than they are on Earth, because the atmosphere is one hundred times thinner than our planet's, and Mars does not have a global magnetic field. As a result many more fast-moving charged particles reach the planet's surface. It is unlikely that life can survive the radiation levels there, though not impossible, because highly radiation-resistant terrestrial organisms exist on Earth, such as *Deinococcus radiodurans*. It is more probable that life would take refuge underground, so samples will have to be collected from under at least a meter or more of soil, where organisms would be protected.

The target of the search for Martian life would shift if there proved to be no living cells in the subsurface or deep subsurface. (Life could thrive deeper underground on Mars than on Earth, due to the more shallow temperature gradient and cooler surface.) The next step would be to investigate if DNA was preserved in the ice, though there is a limit to how long DNA can survive intact. A study by Morten Erik Allentoft, at the University of Copenhagen, in Denmark, suggests that

DNA has a half-life of around half a millennium (521 years), which means that after some five hundred years something like half of the bonds between nucleotides in the backbone of a DNA sample would have broken, and after another five hundred or so years half of the remaining bonds would have been damaged as well, and so on. The current evidence suggests that DNA has a maximum lifetime of around 1.5 million years when held at ideal preservation temperatures,[13] though it is possible that the dry, cold conditions on Mars might have enabled it to last longer.

But, as underscored by the fact that scientists are still arguing over the significance of *Viking* data collected in the 1970s, the best hope we have of detecting life on Mars is gathering direct evidence of it. Ever since Apollo 11 returned the first extraterrestrial samples in the form of forty-nine pounds of moon rock to Earth, it has been a fervent hope that samples of soil from Mars could be obtained for study on Earth. The argument has been that scientists on Earth can make a more thorough and detailed analysis of samples than would be possible by robotic means on the planet itself. While a manned Mars mission remains a distant prospect, we can use machines. The Soviet Union pioneered the use of sample-return robots, notably with Luna 16, which returned 101 grams of material from the moon. In 1975 the Soviets had also planned the very first Martian sample-return project, a twenty-ton robotic mission known as the Mars 5NM, but it was canceled.

Since then extraterrestrial material has been delivered by the *Genesis* mission, which was able to return solar-wind samples to Earth (though it crash-landed in the Utah desert in 2004); the *Stardust* spacecraft, which obtained samples of comet Wild 2 in 2006; and Japan's *Hayabusa* probe, which collected samples after a rendezvous with (and a landing on) asteroid 25143 Itokawa. However, such missions have been fraught with difficulties. For example, Russia's Fobos-Grunt mission to return samples from the Martian moon Phobos failed to leave Earth orbit and crashed into the South Pacific. NASA has long planned a Martian sample-return mission but has yet to secure the funding to get the idea much beyond the drawing board.[14]

Any mission to Mars faces extraordinary technical challenges. Look through the record of space exploration and you'll find that the Red Planet is the Bermuda Triangle of the solar system. It has seen many failed missions, from the 1960s Soviet Mars 1M program (dubbed "Marsnik" in Western media) to Britain's ill-fated *Beagle 2*, which was lost after it left its mother ship to bounce down on the Martian surface in 2003.[15] A successful sample return would mean that the mission craft would have to launch safely; land safely; retrieve a sample from a promising site where water has been present, and preferably several sites; and then return these specimens to Earth. In one such scheme, a mission to collect five-hundred-gram samples from two sites would require fifteen different vehicles and spacecraft and two launch vehicles and would take around three years from launch to return its precious cargo to Earth.[16]

Along the way, steps would have to be taken to ensure that the samples would not be contaminated with terrestrial organisms, although it is highly likely that we have already infected Mars with such organisms after our many missions there. Any parts of the sample-return spacecraft that came into contact with the Martian specimens would have to be sterile to avoid compromising the life-detection experiments. Sequencing machines are now so sensitive that if a single Earth microbe ended up in a sample returned from Mars, it might well ruin the experiment. Contamination has been the bane of many experiments, whether in forensic science or attempts to recover ancient DNA.

Concerns about contamination work both ways. Steps would have to be taken to ensure that any possible Martian life form did not contaminate Earth. (As described above, we are probably a billion years or so too late to worry about this, as they are likely already here.) A Mars sample-return mission would need to comply with planetary-protection requirements more demanding than those for any mission flown to date. The Treaty on Principles Governing the Activities of States in the Exploration and Use of Outer Space, including the Moon and Other Celestial Bodies (or Outer Space Treaty) of 1967[17] states in Article IX, "States parties shall pursue studies of outer space, including the

Moon and other celestial bodies, and conduct exploration of them so as to avoid their harmful contamination and also adverse changes in the environment of the Earth resulting from the introduction of extra-terrestrial matter and, when necessary, adopt appropriate measures for this purpose."

While there are no scientific data to support such concerns, some believe there is good reason to be cautious, based in part, I think, on fear of the unknown, best exemplified by the modern Mary Shelley, the late Michael Crichton. The M.D. turned science-fiction writer was a great storyteller whose books were enjoyable but, like *Frankenstein*, also contained strong anti-science themes, with a blend of fantasy, violence, and retribution of the kind seen in the Brothers Grimm's cautionary fairy tales—"Cinderella," "Red Riding Hood," "Rapunzel," and others—that tapped into the public's deepest fears. In the classic 1971 sci-fi movie *The Andromeda Strain*, a military satellite crashes in the desert, and before it can be retrieved, the inhabitants of a nearby town are decimated by a deadly plague, the eponymous strain that turns out to be quite unlike any life on Earth.[18] Modern science can be applied to circumvent most of the potential problems of sample return from distant celestial bodies.

The latest Mars mission—the *Curiosity* rover, which landed in the crater Gale on August 6, 2012—carries an array of complex instruments, including an alpha-particle X-ray spectrometer; an X-ray diffraction and X-ray fluorescence analyzer; a pulsed neutron source and detector for measuring hydrogen or ice and water; an environmental monitoring station; and an instrument suite that can distinguish between geochemical and biological origin and analyze organics and gases, including oxygen and carbon isotope ratios in carbon dioxide and methane from both atmospheric and solid samples.

Most of these instruments are far more complex than some modern DNA sequencers, such as one made by the company Life Technologies, which can fit on a desktop. This "ion torrent" sequencer uses complementary metal-oxide semiconductor technology, similar to that found in digital cameras, to create the world's smallest solid-state pH meter to translate chemical information into digital information. It employs semiconductor chips, not much bigger than a thumb, with 165 million to 660 million wells, which allow sequencing to be done in parallel. Single-stranded DNA is bound by one end to tiny beads, which are distributed among the minute wells. The wells are subsequently flooded with a solution containing each of the four nucleotides and DNA polymerase. If a nucleotide, for example an A, is added to a DNA template and is then incorporated into the strand of DNA, a single proton (hydrogen ion) is released, causing a pH change in the well, which is detected by the chip. The computer registers which wells had a pH change and records the letter A. This process can be repeated over and over to read a few hundred letters of DNA code in each of the hundreds of millions of wells. Unlike most DNA sequencing technologies, no optics are required to read the signal, so the technique is robust and unaffected by motion. The technology can be made even smaller, which is handy for space missions, where the weight and size of a payload are critical. While there are several issues to be overcome with the sample acquisition, DNA extraction, and preparation for DNA sequencing, none of these represent insurmountable hurdles.

The day is not far off when we will be able send a robotically controlled genome-sequencing unit in a probe to other planets to read the DNA sequence of any alien microbe life that may be there, whether it is still living or preserved. I believe it will be a much greater challenge for NASA or private groups to get an appropriate drill that can extend deep enough to reach the liquid-water level. The good news is that an upcoming mission will be able to drill down a few meters, which could well be sufficient to detect possible signs of frozen life.

It doesn't require a great leap to think that, if Martian microbes are DNA-based, and if we can obtain genome sequences from microbes on Mars and beam them back to Earth, that we should be able to reconstruct the genome. The synthetic version of the Martian genome could then be used to re-create Martian life for detailed studies without having to deal with the incredible logistics of actually bringing the sample back intact. We can rebuild the Martians in a P4 spacesuit lab—that is, a maximum-containment laboratory—instead of risking them splashing down in the ocean or crash-landing in the Amazon. If this process can work from Mars, then we will have a new means of exploring the universe and the hundreds of thousands of Earths and Super-Earths being discovered by the *Kepler* space observatory. To get a sequencer to them soon is out of the question with present-day rocket technology—the planets orbiting the red dwarf Gliese 581 are "only" about twenty-two light-years away, some 1.3×10^{14} miles—but it would take only twenty-two years to get beamed data back, and if advanced life does exist in that system, perhaps it has already been broadcasting sequence information, just as we have done in recent years.

The ability to send DNA software in the form of light will have any number of intriguing ramifications. In the past decade, since my own genome was sequenced, my software has been broadcast in the form of electromagnetic waves, carrying my genetic information far beyond Earth, as they ripple out into space. Borne upon those waves, my life now moves at the speed of light. Whether there is any life form out there capable of making sense of the instructions in my genome is yet another startling thought that spins out of that little question posed by Schrödinger half a century or more ago.

As I concluded my Schrödinger lecture on that warm Dublin evening, I reminded the audience of the incredible journey undertaken by science since Schrödinger himself had mused about the nature of life in his landmark lectures. In the intervening seventy years or so, we have advanced from not knowing the identity of our genetic material to learning that the medium of its message is DNA, to cracking the genetic code, to sequencing genomes, and now to writing genomes to create new life. I have only touched on the opportunities that now beckon as a result of the new knowledge and power that come from the proof by synthesis that DNA is the software of life. We are still riding the powerful waves sent out by Schrödinger's lectures. It is hard to imagine where they will take us in the next seventy years, but wherever this new era of biology is heading, I know that the voyage will be as empowering as it is extraordinary.

Acknowledgments

The great French physiologist Claude Bernard (1813–1878) famously wrote that "Art is I; science is We." That's more true today than ever before. During the past decades I have had many adventures in science and all of them have relied on the efforts of many talented people. I have been able to directly benefit from their wisdom, ingenuity, and creativity, and indirectly from the contributions of the generations of great researchers who came before them, as I first studied proteins and then read, interpreted, and rewrote the basic software of life.

Even with the best will in the world, constraints of space, memory, and time mean that I'll never be able to list everyone who has contributed directly and indirectly to my exploits, let alone explain their contribution in full. Nonetheless, I hope that this volume will at least give a glimpse of that great cooperative endeavor we call science, providing insights into the part dedicated to understanding the most fundamental mechanisms of life itself.

By the same token, this book would not itself have been possible without the help of many others. So I hope you'll indulge me if I single out a few. There is my wife and longtime publicist and partner, Heather Kowalski, who has stood by my side during many turbulent times, from extraordinary highs to bruising lows. For her incredible support and encouragement, I give my heartfelt gratitude. A special mention must be made of my great scientific collaborators, Ham Smith, Clyde Hutchison, and Dan Gibson. I can always rely on them to be diligent, creative, and ingenious in the laboratory. Also, Erling Norrby deserves special thanks for his friendship and long discussions at sea on the nature of life. When it comes to reading draft versions of this particular manuscript, they have been generous with their time and advice. I would also like to single out Lisa Berning and Michelle Tull for their tireless efforts in guiding me through each day.

I would like to thank my agent, John Brockman, for his advice and friendship, and editor Rick Kot, along with his talented colleagues at Viking: we all worked together on my autobiography, *A Life Decoded*, and it has been a pleasure to repeat the experience.

Similarly, I have once again been aided and abetted by Roger Highfield, as my external editor, who in addition to his helpful edits contributed some valuable research. In turn, Roger has benefited from the advice of his colleagues at the Science Museum, London (notably Robert Bud, Peter Morris, and Andrew Nahum), and checked sections with Peter Coveney, Masaru Tomita, and Markus Covert.

Although I have been working on this project for some time, the spark of inspiration that helped me organize my thoughts and drafts into a book was provided by the What Is Life? event, held in 2012 in Dublin as part of the Euroscience Open Forum. I would like to thank Patrick Cunningham, Eamonn Cahill, Luke Drury, David McConnell, Brendan Loftus, Pauric Dempsey, and the Royal Irish Academy for giving me the opportunity to follow in the huge footsteps of Schrödinger, which turned out to be an honor and a pleasure, too. I will not forget how, in the small hours after my lecture, Luke O'Neill treated me, Heather, Erling, and Roger to an impromptu piano concert in the Clarence Hotel.

Notes

Chapter 1

1. Erwin Schrödinger. *What Is Life?* (Cambridge University Press, reprint edition 2012).
2. Thank you, Patrick Cunningham, former chief scientific adviser to the government of Ireland, for pointing this out.
3. Walter J. Moore. *Schrödinger: Life and Thought* (Cambridge: Cambridge University Press, 1989), p. 66.
4. Nikolai V. Timoféeff-Ressovsky, Karl G. Zimmer, and Max Delbrück. "Über die Natur der Genmutation und der Genstruktur" [On the Nature of Gene Mutation and Gene Structure]. Nachrichten von der Gesellschaft der Wissenschaften zu Göttingen, Mathematisch-physikalische Klasse, Fachgruppe VI, Biologie, *Neue Folge* 1, no. 13 (1935): pp. 189–245.
5. Richard Dawkins. *River Out of Eden* (New York: Basic Books, 1995).
6. Motoo Kimura. "Natural selection as the process of accumulating genetic information in adaptive evolution." *Genetical Research* 2 (1961): pp. 127–40.
7. Sydney Brenner. "Life's code script." *Nature* 482 (February 23, 2012): p. 461.
8. W. J. Kress and D. L. Erickson. "DNA barcodes: Genes, genomics, and bioinformatics." *Proceedings of the National Academy of Sciences* 105, no. 8 (2008): pp. 2761–62.
9. Lulu Qian and Erik Winfree. "Scaling up digital circuit computation with DNA strand displacement cascades." *Science* 332, no. 6034 (June 3, 2011): pp. 1196–201.
10. George M. Church, Yuan Gao, and Sriram Kosuri. "Next-generation digital information storage in DNA." *Science* 337, no. 6102 (September 28, 2012): p. 1628.
11. Accessible at http://edge.org/conversation/what-is-life.

Chapter 2

1. Steven Benner. *Life, the Universe . . . and the Scientific Method* (Gainesville, FL: Foundation for Applied Molecular Evolution, 2009), p. 45.
2. Jacques Loeb. *The Dynamics of Living Matter* (New York: Columbia University Press, 1906). Accessible online at http://archive.org/stream/dynamicslivingmooloebgoog#page/n6/mode/2up.
3. Rebecca Lemov. *World as Laboratory: Experiments with Mice, Mazes, and Men* (New York: Hill and Wang, 2005).

4. The book was published in 1627, one year after Bacon died.

5. The original version reads "Salomon."

6. Francis Bacon. *The New Atlantis* (1627). Accessible online at http://oregonstate.edu/instruct/phl302/texts/bacon/atlantis.html.

7. Félix Alexandre Le Dantec. *The Nature and Origin of Life*, trans. Stoddard Dewey (New York: A. S. Barnes, 1906).

8. Some describe Berzelius as a vitalist, but, as John H. Brooke points out, this claim must be treated with care. John H. Brooke. "Wöhler's urea and its vital force: A verdict from the chemists." *Ambix* 15 (1968): pp. 84–114.

9. Quotes from Johannes Büttner, ed. *The Life and Work of Friedrich Wöhler (1800–1882)*, vol. 2., Edition Lewicki-Büttner, edited by Robin Keen (Nordhausen: Verlag T. Bautz, 2005). Digital edition.

10. " . . . denn ich kann, so zu sagen, mein chemisches Wasser nicht halten und muß Ihnen sagen, daß ich Harnstoff machen kann, ohne dazu Nieren oder überhaupt ein Thier, sey es Mensch oder Hund, nöthig zu haben." Otto Wallach, ed. *Briefwechsel zwischen J. Berzelius und F. Wöhler*, vol. 1 (Leipzig: Engelmann, 1901), p. 206.

11. Büttner, *The Life and Work of Friedrich Wöhler (1800–1882)*.

12. Text accessible online at www.biodiversitylibrary.org/item/46624#page/20/mode/1up.

13. Peter J. Ramberg. "The death of vitalism and the birth of organic chemistry: Wöhler's urea synthesis and the disciplinary identity of organic chemistry." *Ambix* 47, no. 3 (November 2000): p. 174.

14. Hermann Kolbe. *Lehrbuch der organischen Chemie* (1854). Accessible online at http://archive.org/stream/ausfhrlicheslehookolbgoog#page/n8/mode/2up.

15. "Friedrich Wöhler obituary." *Scientific American Supplement*, no. 362 (December, 1882). Text is accessible online at www.fullbooks.com/Scientific-American-Supplement-No-3621.html.

16. Brooke, "Wöhler's urea and its vital force."

17. Ibid.

18. John Waller. *Fabulous Science: Fact and Fiction in the History of Scientific Discovery* (Oxford: Oxford University Press, 2010), p. 18.

19. Le Dantec, *The Nature and Origin of Life.*

20. A. M. Turing. "The chemical basis of morphogenesis." *Philosophical Transactions of the Royal Society of London* B 237 (1952): pp. 37–72.

21. Jonathan D. Moreno. "The First Scientist to 'Play God' Was Not Craig Venter." *Science Progress.* http://scienceprogress.org/2010/05/synbio-ethics/.

22. George Dyson. *Turing's Cathedral: The Origins of the Digital Universe* (London: Allen Lane, 2012), p. 284.

23. Schrödinger, *What Is Life?*, pp. 20–21.

24. Brenner, "Life's code script."

25. J. D. Watson and F. H. Crick. "Molecular structure of nucleic acids: A structure for deoxyribose nucleic acid." *Nature* 171, no. 4356 (April 25, 1953): pp. 737–38.

26. A. M. Turing. "Computing machinery and intelligence." *Mind* 59, no. 236 (October 1950): pp. 433–60. Accessible online at www.loebner.net/Prizef/ TuringArticle.html.

27. Ibid.

28. Mark A. Bedau. "Artificial life: Organization, adaptation and complexity from the bottom up." *Trends in Cognitive Sciences* 7, no. 11 (November 2003): pp. 505–512. Accessible online at http://people.reed.edu/~mab/publications/ papers/BedauTICS03.pdf.

29. Dyson, *Turing's Cathedral*, p. 3.

30. George Dyson. "Darwin Among the Machines; or, the Origins of [Artificial] Life." *Edge*. www.edge.org/3rd_culture/dyson/dyson_p2.html.

31. See Charles Ofria and Claus. O Wilke. "Avida: A software platform for research in computational evolutionary biology." *Artificial Life* 10, no. 2 (Spring 2004): pp. 191–229. doi:10.1162/106454604773563612.

32. Dyson, *Turing's Cathedral*, p. 336.

33. George Dyson. "A Universe of Self-Replicating Code." *Edge*, March 26, 2012. http://edge.org/conversation/a-universe-of-self-replicating-code/.

34. Dyson, *Turing's Cathedral*, p. 233.

Chapter 3

1. O. T. Avery. Letter quoted by R. D. Hotchkiss, "Gene, transforming principle, and DNA." In J. Cairns, G. S. Stent, and J. D. Watson, eds. *Phage and the Origins of Molecular Biology* (Cold Spring Harbor, NY: Cold Spring Harbor Laboratory Press, 1966).

2. Erasmus Darwin. *Zoonomia; or the Laws of Organic Life* (1794). Accessible online at http://books.google.co.uk/books?id=A0gSAAAAYAAJ.

3. O. T. Avery, Colin M. MacLeod, and Maclyn McCarty. "Studies on the chemical nature of the substance-inducing transformation of pneumococcal types: Induction of transformation by a desoxyribonucleic acid fraction isolated from pneumococcus type III." *Journal of Experimental Medicine* 79 (January 1944): pp. 137–58.

4. Jacob Stegenga. "The chemical characterization of the gene: Vicissitudes of evidential assessment hist." *History and Philosophy of the Life Sciences* 33 (2011): pp. 105–27.

5. F. Sanger. "The arrangement of amino acids in proteins." *Advances in Protein Chemistry* 7 (1952): pp. 1–66.

6. Antony O. W. Stretton. "The first sequence: Fred Sanger and insulin." *Genetics* 162, no. 2 (October 1, 2002): pp. 527–32.

7. F. Sanger. "Sequences, sequences, and sequences." *Annual Review of Biochemistry* 57 (1988): pp. 1–28.

8. A. D. Hershey and M. Chase. "Independent functions of viral protein and nucleic acid in growth of bacteriophage." *Journal of General Physiology* 36 (1952): pp. 39–56.

9. Photo 51 taken by Franklin is now part of the History of Science collection at the Venter Institute.

10. Watson and Crick, "Molecular structure of nucleic acids," pp. 737–38.

11. http://oralhistories.library.caltech.edu/33/0/OH_Sinsheimer.pdf.

12. Erling Norrby. *Nobel Prizes and Life Sciences* (Singapore: World Scientific Publishing, 2010).

13. D. A. Jackson, R. H. Symons, and P. Berg. "Biochemical method for inserting new genetic information into DNA of simian virus 40: Circular SV40 DNA molecules containing lambda phage genes and the galactose operon of *Escherichia coli*." *Proceedings of the National Academy of Sciences* 69, no. 10 (1972): pp. 2904–9.

14. R. Jaenisch and B. Mintz. "Simian Virus 40 DNA sequences in DNA of healthy adult mice derived from preimplantation blastocysts injected with viral DNA." *Proceedings of the National Academy of Sciences* 71, no. 4 (1974): pp. 1250–54.

15. Joshua Lederberg. "DNA splicing: Will fear rob us of its benefits?" *Prism* 3 (November 1975): pp. 33–37.

16. Robert Hooke. *Micrographia* (1665).

17. Ute Diechmann. " 'Molecular' versus 'colloidal': Controversies in biology and biochemistry, 1900–1940." *Bulletin for the History of Chemistry* 32, no. 2 (2007): pp. 105–118. Accessible online at www.scs.illinois.edu/~mainzv/HIST/awards/OPA%20Papers/2009-Deichmann.pdf.

18. Bruce Alberts. "The cell as a collection of protein machines: Preparing the next generation of molecular biologists." *Cell* 92 (February 6, 1998): p. 291.

19. Marco Piccolino. "Biological machines: From mills to molecules." *Nature Reviews: Molecular Cell Biology* 1 (November 2000): pp. 149–52.

20. Gregory S. Engel, Tessa R. Calhoun, Elizabeth L. Read, Tae-Kyu Ahn, Tomáö Mancal, Yuan-Chung Cheng, Robert E. Blankenship, and Graham R. Fleming. "Evidence for wavelike energy transfer through quantum coherence in photosynthetic systems." *Nature* 446 (April 12, 2007): pp. 782–86.

21. Graham R. Fleming, Gregory D. Scholes, and Yuan-Chung Cheng. "Quantum effects in biology." *Procedia Chemistry* 3, no. 1 (2011): pp. 38–57. Accessible online at www.sciencedirect.com/science/article/pii/S1876619611000507.

22. M. A. Martin-Delgado. "On quantum effects in a theory of biological evolution." *Scientific Reports* 2 (March 12, 2012): article number 302.

23. Joachim Frank and Rajendra Kumar Agrawal. "A ratchet-like inter-subunit reorganization of the ribosome during translocation." *Nature* 406 (July 20, 2000): pp. 318–22.

24. An angstrom is about the length of an atom; there are ten billion in a meter.

25. See http://library.cshl.edu/oralhistory/interview/cshl/memories/harry -noller-and-ribosome/.

26. C. Napoli, C. Lemieux, and R. Jorgensen. "Introduction of a chimeric chalcone synthase gene into petunia results in reversible co-suppression of homologous genes in trans." *Plant Cell* 2, no. 4 (1990): pp. 279–89.

27. E. Eden, N. Geva-Zatorsky, I. Issaeva, A. Cohen, E. Dekel, T. Danon, L. Cohen, A. Mayo, and U. Alon. "Proteome half-life dynamics in living human cells." *Science* 331, no. 6018 (February 11, 2011): pp. 764–68.

28. The protein-folding video can be viewed at www.youtube.com/watch?v =sD6vyfTtE4U&feature=youtu.be. See also www.ks.uiuc.edu/Research/ folding/.

29. Fei Sun, Zhibao Mi, Steven B. Condliffe, Carol A. Bertrand, Xiaoyan Gong, Xiaoli Lu, Ruilin Zhang, Joseph D. Latoche, Joseph M. Pilewski, Paul D. Robbins, and Raymond A. Frizzell. "Chaperone displacement from mutant cystic fibrosis transmembrane conductance regulator restores its function in human airway epithelia." *FASEB Journal* 22, no. 9 (September 2, 2008): pp. 3255–63.

30. Varshavsky Alexander. "The N-end rule pathway of protein degradation." *Genes to Cells* 2, no. 1 (1997): pp. 13–28.

31. George Oster and Hongyun Wang. "How Protein Motors Convert Chemical Energy into Mechanical Work." In M. Schliwa, ed., *Molecular Motors* (Weinheim, Germany: Wiley-VCH, 2003). Accessible online at http://users .soe.ucsc.edu/~hongwang/publications/Schliwa_08.pdf

32. Ibid.

Chapter 4

1. Sydney Brenner. "Biochemistry Strikes Back." In Jan Witkowski, ed. *The Inside Story: DNA to RNA to Protein* (Cold Spring Harbor, NY: Cold Spring Harbor Laboratory Press, 2005), p. 367.

2. Walter Gilbert and Allan Maxam. "The nucleotide sequence of the lac operator." *Proceedings of the National Academy of Sciences* 70, no. 12, part 1 (1973): pp. 3581–84.

3. R. W. Holley, G. A. Everett, J. T. Madison, and A. Zamir. "Nucleotide sequences in the yeast alanine transfer ribonucleic acid." *Journal of Biological Chemistry* 240, no. 5 (May 1965): pp. 2122–28.

4. G. G. Brownlee, F. Sanger, and B. G. Barrell. "Nucleotide sequence of 5S -ribosomal RNA from *Escherichia coli*." *Nature* 215, no. 5102 (1967): pp. 735–36.

5. F. Sanger, G. M. Air, B. G. Barrell, N. L. Brown, A. R. Coulson, C. A. Fiddes, C. A. Hutchinson, P. M. Slocombe, et al. "Nucleotide sequence of bacteriophage φX174 DNA." *Nature* 265, no. 5596 (1977): pp. 687–95.

6. http://oralhistories.library.caltech.edu/33.

7. J. Craig Venter. *A Life Decoded: My Genome: My Life* (New York: Viking, 2007).

8. Frederick Sanger, Nobel Lecture, December 8, 1980.

9. F. Z. Chung, K. U. Lentes, J. Gocayne, M. Fitzgerald, D. Robinson, A. R. Kerlavage, C. M. Fraser, and J. C. Venter. "Cloning and sequence analysis of the human brain beta-adrenergic receptor: Evolutionary relationship to rodent and avian beta-receptors and porcine muscarinic receptors." *FEBS Letters* 211, no. 2 (January 26, 1987): pp. 200–206.

10. Lloyd M. Smith, Jane Z. Sanders, Robert J. Kaiser, Peter Hughes, Chris Dodd, Charles R. Connell, Cheryl Heiner, Stephen B. H. Kent, and Leroy E. Hood. "Fluorescence detection in automated DNA sequence analysis." *Nature* 321 (June 12, 1986): pp. 674–79.

11. M. D. Adams, J. M. Kelley, J. D. Gocayne, M. Dubnick, M. H. Polymeropoulos, H. Xiao, C. R. Merril, A. Wu, B. Olde, R. F. Moreno, et al. "Complementary DNA sequencing: Expressed sequence tags and human genome project." *Science* 252 (1991): pp. 1651–56.

12. Venter, *A Life Decoded*.

13. Ibid.

14. Ibid.

15. C. M. Fraser, J. D. Gocayne, O. White, M. D. Adams, R. A. Clayton, R. Fleischmann, C. J. Bult, A. R. Kerlavage, G. Sutton, J. M. Kelley, J. L. Fritchman, J. F. Weidman, K. V. Small, M. Sandusky, J. Fuhrmann, D. Nguyen, T. R. Utterback, D. M. Saudek, C. A. Phillips, J. M. Merrick, J. Tomb, B. A. Dougherty, K. F. Bott, P. Hu, T. S. Lucier, S. N. Peterson, H. O. Smith, C. A. Hutchison, J. C. Venter. "The minimal gene complement of *Mycoplasma genitalium*." *Science* 270 (1995): pp. 397–403.

16. M. G. Lee and P. Nurse. "Complementation used to clone a human homologue of the fission yeast cell cycle control gene cdc2." *Nature* 327 (1987): pp. 31–35.
17. Eugene V. Koonin, Arcady R. Mushegian, and Kenneth E. Rudd. "Sequencing and analysis of bacterial genomes." *Current Biology* 6, no. 4 (1996): pp. 404–16.
18. Carl R. Woese and George E. Fox. "Phylogenetic structure of the prokaryotic domain: The primary kingdoms." *Proceedings of the National Academy of Sciences* 74, no. 11 (1977): pp. 5088–90.
19. C. J. Bult, O. White, G. J. Olsen, L. Zhou, R. D. Fleischmann, G. G. Sutton, J. A. Blake, L. M. FitzGerald, R. A. Clayton, J. D. Gocayne, A. R. Kerlavage, B. A. Dougherty, J. F. Tomb, M. D. Adams, C. I. Reich, R. Overbeek, E. F. Kirkness, K. G. Weinstock, J. M. Merrick, A. Glodek, J. L. Scott, S. M. Geoghagen, J. F. Weidman, J. L. Fuhrmann, D. Nguyen, T. R. Utterback, J. M. Kelley, J. D. Peterson, P. W. Sadow, M. C. Hanna, M. D. Cotton, K. M. Roberts, M. A. Hurst, B. P. Kaine, M. Borodovsky, H. P. Klenk, C. M. Fraser, H. O. Smith, C. R. Woese, and J. C. Venter. "Complete genome sequence of the methanogenic archaeon, *Methanococcus jannaschii*." *Science* 372 (1996): pp. 1058–73.
20. Venter, *A Life Decoded*.
21. Simonetta Gribaldo, Anthony M. Poole, Vincent Daubin, Patrick Forterre, and Céline Brochier-Armanet. "The origin of eukaryotes and their relationship with the Archaea: Are we at a phylogenomic impasse?" *Nature Reviews: Microbiology* 8 (October 2010): pp. 743–52. doi:10.1038/nrmicro2426.
22. D. Raoult, M. Drancourt, S. Azza, et al. "Nanobacteria are mineralo fetuin complexes." *PLOS Pathogens* 4, no. 2 (February 2008): e41. J. M. García-Ruiz, E. Melero-García, and S. T. Hyde. "Morphogenesis of self-assembled nanocrystalline materials of barium carbonate and silica." *Science* 323, no. 5912 (January 2009): pp. 362–65.
23. J.-F. Tomb, O. White, A. R. Kerlavage, R. A. Clayton, G. G. Sutton, R. D. Fleischmann, K. A. Ketchum, H.-P. Klenk, S. Gill, B. A. Dougherty, K. Nelson, J. Quackenbush, L. Zhou, E. F. Kirkness, S. Peterson, B. Loftus, D. Richardson, R. Dodson, H. G. Khalak, A. Glodek, K. McKenney, L. M. Fitzgerald, N. Lee, M. D. Adams, E. K. Hickey, D. E. Berg, J. D. Gocayne, T. R. Utterback, J. D. Peterson, J. M. Kelley, M. D. Cotton, J. M. Weidman, C. Fujii, C. Bowman, L. Whatthey, E. Wallin, W. S. Hayes, M. Borodovsky, P. D. Karp, H. O. Smith, C. M. Fraser, and J. C. Venter. "The complete genome sequence of the gastric pathogen *Helicobacter pylori*." *Nature* 388 (1997): pp. 539–47.
24. See www.nobelprize.org/nobel_prizes/medicine/laureates/2005/marshall -cv.html.
25. H. P. Klenk, R. A. Clayton, J. F. Tomb, O. White, K. E. Nelson, K. A. Ketchum, R. J. Dodson, M. Gwinn, E. K. Hickey, J. D. Peterson, D. L. Richardson, A. R.

Kerlavage, D. E. Graham, N. C. Kyrpides, R. D. Fleischmann, J. Quackenbush, N. H. Lee, G. G. Sutton, S. Gill, E. F. Kirkness, B. A. Dougherty, K. McKenney, M. D. Adams, B. Loftus, S. Peterson, C. I. Reich, L. K. McNeil, J. H. Badger, A. Glodek, L. Zhou, R. Overbeek, J. D. Gocayne, J. F. Weidman, L. McDonald, T. Utterback, M. D. Cotton, T. Spriggs, P. Artiach, B. P. Kaine, S. M. Sykes, P. W. Sadow, K. P. D'Andrea, C. Bowman, C. Fujii, S. A. Garland, T. M. Mason, G. J. Olsen, C. M. Fraser, H. O. Smith, C. R. Woese, and J. C. Venter. "The complete genome sequence of the hyperthermophilic, sulphate-reducing archaeon *Archaeoglobus fulgidu*." *Nature* 390 (1997): pp. 364–70.

26. A. Goffeau, B. G. Barrell, H. Bussey, R. W. Davis, B. Dujon, H. Feldmann, F. Galibert, J. D. Hoheisel, C. Jacq, M. Johnston, E. J. Louis, H. W. Mewes, Y. Murakami, P. Philippsen, H. Tettelin, and S. G. Oliver. "Life with 6000 genes." *Science* 274, no. 5287 (1996): pp. 546, 563–67.

27. See www.nobelprize.org/nobel_prizes/medicine/laureates/1983/press.html.

Chapter 5

1. Text accessible at www.presidency.ucsb.edu/ws/index.php?pid=28606.

2. Sinsheimer: "The two smallest bacterial viruses one could find in the literature were one called S13, which had been discovered in England, and Phi X 174, which had been discovered in France. And you might say, 'Well, how were they discovered?' Well, people were just sort of categorizing viruses. They would take some sewage from the Paris sewers and find how many viruses they could find and which hosts, or cells, they would plate on and what the plaques looked like. The name Phi X 174 means it was the 174th virus in the tenth series of phages that they got; that's all. No more meaning than that. This was the tenth set of experiments they'd done—X was actually ten." http://oralhistories.library.caltech.edu/33.

3. Text accessible at www.presidency.ucsb.edu/ws/index .php?pid=28606#axzz1ufDunxa6.

4. "Creating Life in the Test Tube." Arthur Kornberg Papers, 1959–1970. National Library of Medicine. Kornberg DNA synthesis references: "Closer to Synthetic Life." *Time*, December 22, 1967, p. 66; "Viable Synthetic DNA." *Science News*, December 30, 1967, pp. 629–30.

5. S. Nagata, H. Taira, A. Hall, L. Johnsrud, M. Streuli, J. Ecsödi, et al. "Synthesis in *E. coli* of a polypeptide with human leukocyte interferon activity." *Nature* 284 (1980): pp. 316–20.

6. M. A. Billeter, J. E. Dahlberg, H. M. Goodman, J. Hindley, and C. Weissmann. "Sequence of the first 175 nucleotides from the 5 terminus of Qbeta RNA synthesized in vitro." *Nature* 224 (1969): pp. 1083–86.

7. Charles Weissmann. "End of the road." *Prion* 6, no. 2 (April 1, 2012): pp. 97–104. doi:10.4161/pri.19778.

8. T. Taniguchi, M. Palmieri, and C. Weissmann. "QB DNA-containing hybrid plasmids giving rise to QB phage formation in the bacterial host." *Nature* 274 (1978): pp. 223–28.

9. V. R. Racaniello and D. Baltimore. "Molecular cloning of poliovirus cDNA and determination of the complete nucleotide sequence of the viral genome." *Proceedings of the National Academy of Sciences* 78 (1981): pp. 4887–91.

10. See footnote nine in Keril J. Blight, Alexander A. Kolykhalov, and Charles M. Rice. "Efficient initiation of HCV RNA replication in cell culture." *Science* 290, no. 5498 (December 8, 2000): pp. 1972–74.

11. Eckard Wimmer, Steffen Mueller, Terrence M. Tumpey, and Jeffery K. Taubenberger. "Synthetic viruses: A new opportunity to understand and prevent viral disease." *Nature Biotechnology* 27, no. 12 (December 2009): p. 1163.

12. J. Craig Venter, Karin Remington, John F. Heidelberg, Aaron L. Halpern, Doug Rusch, Jonathan A. Eisen, Dongying Wu, Ian Paulsen, Karen E. Nelson, William Nelson, Derrick E. Fouts, Samuel Levy, Anthony H. Knap, Michael W. Lomas, Ken Nealson, Owen White, Jeremy Peterson, Jeff Hoffman, Rachel Parsons, Holly Baden-Tillson, Cynthia Pfannkoch, Yu-Hui Rogers, and Hamilton O. Smith. "Environmental genome shotgun sequencing of the Sargasso Sea." *Science* 304, no. 5667 (April 2, 2004): pp. 66–74.

13. http://oralhistories.library.caltech.edu/33/0/OH_Sinsheimer.pdf.

14. Walter Fiers and Robert L. Sinsheimer. "The structure of the DNA of bacteriophage [[phi]]X174, III: Ultracentrifugal evidence for a ring structure." *Journal of Molecular Biology* 5, no. 4 (October 1962): pp. 424–434.

15. Jeronimo Cello, Aniko V. Paul, and Eckard Wimmer. "Chemical synthesis of poliovirus cDNA: Generation of infectious virus in the absence of natural template." *Science* 297, no. 5583 (August 9, 2002): pp. 1016–18. doi:10.1126/science.1072266.

16. See www.fbi.gov/about-us/history/famous-cases/anthrax-amerithrax/amerithrax-investigation.

17. See www.umass.edu/legal/derrico/amherst/lord_jeff.html.

18. http://www.who.int/csr/disease/smallpox/en/index.html.

19. Mildred K. Cho, David Magnus, Arthur L. Caplan, Glenn McGee, and the Ethics of Genomics Group. "Ethical considerations in synthesizing a minimal genome." *Science* 286, no. 5447 (December 10, 1999): pp. 2087–90.

20. George Church and Ed Regis. *Regenesis: How Synthetic Biology Will Reinvent Nature and Ourselves* (New York: Basic Books, 2012), p. 9.

21. Cho, et. al, "Ethical Considerations in Synthesizing a Minimal Genome."

22. Kenneth I. Berns, Arturo Casadevall, Murray L. Cohen, Susan A. Ehrlich, Lynn W. Enquist, J. Patrick Fitch, David R. Franz, Claire M. Fraser-Liggett, Christine M. Grant, Michael J. Imperiale, Joseph Kanabrocki, Paul S. Keim, Stanley M. Lemon, Stuart B. Levy, John R. Lumpkin, Jeffery F. Miller, Randall Murch, Mark E. Nance, Michael T. Osterholm, David A. Relman, James A. Roth, and Anne K. Vidaver. "Policy: Adaptations of avian flu virus are a cause for concern." *Nature* 482 (February 9, 2012): pp. 153–154. doi:10.1038/482153a.

Chapter 6

1. Samuel Butler. *Erewhon* (1872), pp. 318–19.
2. Mirel, or polyhydroxybutyrate.
3. See www.scarabgenomics.com/.
4. Farren J. Isaacs, Peter A. Carr, Harris H. Wang, Marc J. Lajoie, Bram Sterling, Laurens Kraal, Andrew C. Tolonen, Tara A. Gianoulis, Daniel B. Goodman, Nikos B. Reppas, Christopher J. Emig, Duhee Bang, Samuel J. Hwang, Michael C. Jewett, Joseph M. Jacobson, and George M. Church. "Precise manipulation of chromosomes in vivo enables genome-wide codon replacement." *Science* 333, no. 6040 (July 15, 2011): pp. 348–53.
5. Tae Seok Moon, Chunbo Lou, Alvin Tamsir, Brynne C. Stanton, and Christopher A. Voigt. "Genetic programs constructed from layered logic gates in single cells." *Nature* 491 (November 8, 2012): pp. 249–53.
6. Piro Siuti, John Yazbek, and Timothy K. Lu. "Synthetic circuits integrating logic and memory in living cells." *Nature Biotechnology* 31 (2013): pp. 448–452. doi:10.1038/nbt.2510.
7. S. Wuchty, B. F. Jones, and B. Uzzi. "The increasing dominance of teams in production of knowledge." *Science* 316, no. 5827 (2007): pp. 1036–39.
8. Kira J. Weissman and Peter F. Leadlay. "Combinatorial biosynthesis of reduced polyketides." *Nature Reviews: Microbiology* 3 (December 2005): 925–36.
9. Richard E. Green, Johannes Krause, Adrian W. Briggs, Tomislav Maricic, Udo Stenzel, Martin Kircher, Nick Patterson, Heng Li, Weiwei Zhai, Markus Hsi-Yang Fritz, Nancy F. Hansen, Eric Y. Durand, Anna-Sapfo Malaspinas, Jeffrey D. Jensen, Tomas Marques-Bonet, Can Alkan, Kay Prüfer, Matthias Meyer, Hernán A. Burbano, Jeffrey M. Good, Rigo Schultz, Ayinuer Aximu-Petri, Anne Butthof, Barbara Höber, Barbara Höffner, Madlen Siegemund, Antje Weihmann, Chad Nusbaum, Eric S. Lander, Carsten Russ, Nathaniel Novod, Jason Affourtit, Michael Egholm, Christine Verna, Pavao Rudan, Dejana Brajkovic, Zeljko Kucan, Ivan Guöic, Vladimir B. Doronichev, Liubov V. Golovanova, Carles Lalueza-Fox, Marco de la Rasilla, Javier Fortea, Antonio Rosas, Ralf W. Schmitz, Philip L. F. Johnson, Evan E. Eichler, Daniel Falush,

Ewan Birney, James C. Mullikin, Montgomery Slatkin, Rasmus Nielsen, Janet Kelso, Michael Lachmann, David Reich, and Svante Pääbo. "A draft sequence of the Neandertal genome." *Science* 328, no. 5979 (May 7, 2010): pp. 710–22.

10. See http://mammoth.psu.edu/index.html.

11. Church and Regis, *Regenesis*, p. 11.

12. O. White, J. A. Eisen, J. F. Heidelberg, E. K. Hickey, J. D. Peterson, R. J. Dodson, D. H. Haft, M. L. Gwinn, W. C. Nelson, D. L. Richardson, K. S. Moffat, H. Qin, L. Jiang, W. Pamphile, M. Crosby, M. Shen, J. J. Vamathevan, P. Lam, L. McDonald, T. Utterback, C. Zalewski, K. S. Makarova, L. Aravind, M. J. Daly, K. W. Minton, R. D. Fleischmann, K. A. Ketchum, K. E. Nelson, S. Salzberg, H. O. Smith, J. C. Venter, and C. M. Fraser. "Complete genome sequencing of the radioresistant bacterium, *Deinococcus radiodurans* R1." *Science* 286, no. 5444 (November 19, 1999): pp. 1571–77.

13. J. C. Venter and C. Yung, eds. *Target-Size Analysis of Membrane Proteins* (New York: Alan R. Liss, 1987).

14. Mitsuhiro Itaya, Kenji Tsuge, Maki Koizumi, and Kyoko Fujita. "Combining two genomes in one cell: Stable cloning of the *Synechocystis* PCC6803 genome in the *Bacillus subtilis* 168 genome." *Proceedings of the National Academy of Sciences* 102, no. 44 (2005): pp. 15971–76. doi:10.1073/pnas.0503868102.

15. V. Larionov, N. Kouprina, J. Graves, X. N. Chen, J. R. Korenberg, and M. A. Resnick. "Specific cloning of human DNA as yeast artificial chromosomes by transformation-associated recombination." *Proceedings of the National Academy of Sciences* 93, no. 1 (1996): pp. 491–96.

Chapter 7

1. Thomas Kuhn. *The Structure of Scientific Revolutions* (Chicago: University of Chicago Press, 1962), pp. 84–85.

2. C. Lartigue, J. I. Glass, N. Alperovich, R. Pieper, P. P. Parmar, C. A. Hutchison III, H. O. Smith, and J. C. Venter. "Genome transplantation in bacteria: Changing one species to another." *Science* 317, no. 5838 (August 3, 2007): pp. 632–38.

3. I. Wilmut, A. E. Schnieke, J. McWhir, A. J. Kind, and K. H. Campbell. "Viable offspring derived from fetal and adult mammalian cells." *Nature* 385, no. 6619 (1997): pp. 810–13.

4. I. Wilmut and R. Highfield. *After Dolly: The Uses and Misuses of Human Cloning* (New York: Norton, 2006).

5. S. M. Willadsen. "Nuclear transplantation in sheep embryos." *Nature* 320 (March 6, 1986): pp. 63–65. Accessible online at www.nature.com/nature/journal/v320/n6057/abs/320063a0.html.

6. H. Spemann. *Embryonic Development and Induction* (New Haven, CT: Yale University Press, 1938).

7. J. B. Gurdon. "The developmental capacity of nuclei taken from intestinal epithelium cells of feeding tadpoles." *Journal of Embryology and Experimental Morphology* 34 (1962): pp. 93–112.

8. See www.nobelprize.org/nobel_prizes/medicine/laureates/2012/press.html#. See also Wilmut and Highfield, *After Dolly*.

9. J. F. Heidelberg, J. A. Eisen, W. C. Nelson, R. A. Clayton, M. L. Gwinn, R. J. Dodson, D. H. Haft, E. K. Hickey, J. D. Peterson, L. Umayam, S. R. Gill, K. E. Nelson, T. D. Read, H. Tettelin, D. Richardson, M. D. Ermolaeva, J. Vamathevan, S. Bass, H. Qin, I. Dragoi, P. Sellers, L. McDonald, T. Utterback, R. D. Fleishmann, W. C. Nierman, O. White, S. L. Salzberg, H. O. Smith, R. R. Colwell, J. J. Mekalanos, J. C. Venter, and C. M. Fraser. "DNA sequence of both chromosomes of the cholera pathogen *Vibrio cholerae.*" *Nature* 406, no. 6795 (August 3, 2000): pp. 477–83.

10. World Health Organisation. "Cholera: Fact Sheet No. 107," August 2011. www .who.int/mediacentre/factsheets/fs107/en/index.html.

11. A. Fischer, B. Shapiro, C. Muriuki, M. Heller, C. Schnee, et al. "The origin of the '*Mycoplasma mycoides* cluster' coincides with domestication of ruminants." *PLOS ONE* 7, no. 4 (2012): e36150.

12. DNA can be in a supercoiled state only if it is completely intact. Any cuts or nicks in the DNA will cause it to uncoil.

13. The idea is thought to date back to at least Laplace, who said, "The weight of evidence for an extraordinary claim must be proportioned to its strangeness."

Chapter 8

1. "In 'The Value of Science,' *What Do You Care What Other People Think?*" (1988, 2001), p. 247. Collected in Richard P. Feynman. *The Pleasure of Finding Things Out* (New York, Perseus, 1999).

2. H. Gardner. *Creating Minds: An Anatomy of Creativity Seen Through the Lives of Freud, Einstein, Picasso, Stravinsky, Eliot, Graham, and Ghandi* (New York: HarperCollins, 1993).

3. See www.nobelprize.org/nobel_prizes/chemistry/laureates/2008/shimomura .html.

4. C. Lartigue, S. Vashee, M. A. Algire, R. Y. Chuang, G. A. Benders, L. Ma, V. N. Noskov, E. A. Denisova, D. G. Gibson, N. Assad-Garcia, N. Alperovich, D. W. Thomas, C. Merryman, C. A. Hutchison III, H. O. Smith, J. C. Venter, and J. I. Glass. "Creating bacterial strains from genomes that have been cloned and engineered in yeast." *Science* 325, no. 5948 (September 25, 2009): pp. 1693–96.

doi:10.1126/science.1173759. Gwynedd A. Benders, Vladimir N. Noskov, Evgeniya A. Denisova, Carole Lartigue, Daniel G. Gibson, Nacyra Assad-Garcia, Ray-Yuan Chuang, William Carrera, Monzia Moodie, Mikkel A. Algire, Quang Phan, Nina Alperovich, Sanjay Vashee, Chuck Merryman, J. Craig Venter, Hamilton O. Smith, John I. Glass, and Clyde A. Hutchison III. "Cloning whole bacterial genomes in yeast." *Nucleic Acids Research* 38, no. 8 (May 2010): pp. 2558–69. doi:10.1093/nar/gkq119. D. G. Gibson, G. A. Benders, K. C. Axelrod, J. Zaveri, M. A. Algire, M. Moodie, M. G. Montague, J. C. Venter, H. O. Smith, and C. A. Hutchison III. "One-step assembly in yeast of 25 overlapping DNA fragments to form a complete synthetic *Mycoplasma genitalium* genome." *Proceedings of the National Academy of Sciences* 105 (2008): pp. 20404–9.

5. Carole Lartigue, Sanjay Vashee, Mikkel A. Algire, Ray-Yuan Chuang, Gwynedd A. Benders, Li Ma, Vladimir N. Noskov, Evgeniya A. Denisova, Daniel G. Gibson, Nacyra Assad-Garcia, Nina Alperovich, David W. Thomas, Chuck Merryman, Clyde A. Hutchison III, Hamilton O. Smith, J. Craig Venter, and John I. Glass. "Creating bacterial strains from genomes that have been cloned and engineered in yeast." *Science* 325, no. 5948 (September 25, 2009): pp. 1693–96.

6. For example, Windows XP has around forty-five million. See www.facebook.com/windows/posts/155741344475532.

7. Daniel G. Gibson, John I. Glass, Carole Lartigue, Vladimir N. Noskov, Ray-Yuan Chuang, Mikkel A. Algire, Gwynedd A. Benders, Michael G. Montague, Li Ma, Monzia M. Moodie, Chuck Merryman, Sanjay Vashee, Radha Krishnakumar, Nacyra Assad-Garcia, Cynthia Andrews-Pfannkoch, Evgeniya A. Denisova, Lei Young, Zhi-Qing Qi, Thomas H. Segall-Shapiro, Christopher H. Calvey, Prashanth P. Parmar, Clyde A. Hutchison III, Hamilton O. Smith, and J. Craig Venter. "Creation of a bacterial cell controlled by a chemically synthesized genome." *Science* 329, no. 5987 (July 2, 2010): pp. 52–56.

Chapter 9

1. Daniel E. Koshland, Jr. "The seven pillars of life." *Science* 295, no. 5563 (March 22, 2002): pp. 2215–16. doi:10.1126/science.1068489.

2. Robert B. Leighton and Richard Feynman. *The Feynman Lectures on Physics*, vol. I, 8–2 (Boston: Addison–Wesley, 1964).

3. Ian Sample. "Craig Venter Creates Synthetic Life Form." *Guardian*, May 20, 2010. www.guardian.co.uk/science/2010/may/20/craig-venter-synthetic-life-form.

4. Phillip F. Schewe. *Maverick Genius: The Pioneering Odyssey of Freeman Dyson* (New York: Thomas Dunne, 2013).

5. Nicholas Wade. "Researchers Say They Created a 'Synthetic Cell.'" *New York Times*, May 20, 2010. www.nytimes.com/2010/05/21/science/21cell.html.

6. See the Leveson Inquiry: www.levesoninquiry.org.uk/.

7. Fiona Macrae. "Scientist Accused of Playing God after Creating Artificial Life by Making Designer Microbe from Scratch—But Could It Wipe Out Humanity?" *Daily Mail*, June 3, 2010. www.dailymail.co.uk/sciencetech/article-1279988/Artificial-life-created-Craig-Venter—wipe-humanity.html.

8. New Directions: The Ethics of Synthetic Biology and Emerging Technologies. Presidential Commission for the Study of Bioethical Issues, Washington D.C., December 2010. www.bioethics.gov.

9. "Vatican greets development of first synthetic cell with caution." *The Catholic Transcript Online*. http://www.catholictranscript.org/about/1429-vatican-greets-development-of-first-synthetic-cell-with-caution.html.

10. Text accessible online at http://life.ou.edu/pubs/fatm/.

11. Paul Nurse. "Wee beasties." *Nature* 432, no. 7017 (December 2004): p. 557.

12. Shao Jun Du, Zhiyuan Gong, Garth L. Fletcher, Margaret A. Shears, Madonna J. King, David R. Idler, and Choy L. Hew. "Growth enhancement in transgenic Atlantic salmon by the use of an 'all fish' chimeric growth hormone gene construct." *Bio/Technology* 10, no. 2 (1992): pp. 176–81.

13. William B. Whitman, David C. Coleman, and William J. Wiebe. "Prokaryotes: The unseen majority." *Proceedings of the National Academy of Sciences* 95, no. 12 (1998): pp. 6578–83. doi:10.1073/pnas.95.12.6578.

14. Francis Crick. *Life Itself: Its Origin and Nature* (New York: Simon and Schuster, 1981).

15. C. C. Price, "The new era in science." *Chemical and Engineering News* 43, no. 39 (1965): p. 90.

16. Stanley L. Miller. "Production of amino acids under possible primitive earth conditions." *Science* 117, no. 3046 (May 1953): pp. 528–29.

17. J. Oró and A. P. Kimball. "Synthesis of purines under possible primitive earth conditions. I. Adenine from hydrogen cyanide." *Archives of Biochemistry and Biophysics* 94 (August 1961): pp. 217–27. J. Oró and S. S. Kama. "Amino-acid synthesis from hydrogen cyanide under possible primitive earth conditions." *Nature* 190, no. 4774 (April 1961): pp. 442–43. J. Oró in S. W. Fox, ed. *Origins of Prebiological Systems and of Their Molecular Matrices* (New York: Academic Press, 1967), p. 137.

18. Thomas R. Cech. "The RNA worlds in context." *Cold Spring Harbor Perspectives in Biology* 4, no. 7 (July 1, 2012). Accessible online at http://cshperspectives.cshlp.org/content/4/7/a006742.

19. Carl Woese. *The Genetic Code* (New York: Harper and Row, 1967).
20. K. Kruger, P. J. Grabowski, A. J. Zaug, J. Sands, D. E. Gottschling, T. R. Cech. "Self-splicing RNA: Autoexcision and autocyclization of the ribosomal RNA intervening sequence of *Tetrahymena.*" *Cell* 31, no. 1 (November 1982): pp. 147–57.
21. C. Guerrier-Takada, K. Gardiner, T. Marsh, N. Pace, and S. Altman. "The RNA moiety of ribonuclease P is the catalytic subunit of the enzyme." *Cell* 35, no. 3, part 2 (1983): pp. 849–57.
22. The 1989 Nobel Prize in Chemistry was awarded to Thomas R. Cech and Sidney Altman "for their discovery of catalytic properties of RNA."
23. For an up-to-date review, I refer readers to David Deamer's *First Life: Discovering the Connections Between Stars, Cells and How Life Began* (Berkeley and Los Angeles: University of California Press, 2011).
24. Jack Szostak was awarded the 2009 Nobel Prize in Physiology or Medicine, along with Elizabeth Blackburn and Carol W. Greider, for the discovery of how chromosomes are protected by telomeres.
25. Jack W. Szostak, David P. Bartel, and P. Luigi Luisi. "Synthesizing life." *Nature* 409 (January 18, 2001): pp. 387–90.
26. Rich Roberts, now at New England Biolabs, shared the 1993 Nobel Prize in Physiology or Medicine with Phillip Sharp for the discovery of introns in eukaryotic DNA and the mechanism of gene splicing.
27. I. A. Chen, R. W. Roberts, and J. W. Szostak. "The emergence of competition between model protocells." *Science* 305 (September 3, 2004): pp. 1474–76.
28. Kurt J. Isselbacher. "Paul C. Zamecnik (1912–2009)." *Science* 326, no. 5958 (December 4, 2009): p. 1359.
29. M. W. Nirenberg and H. J. Matthaei. "The dependence of cell-free protein synthesis in *E. coli* upon naturally occurring or synthetic polyribonucleotides." *Proceedings of the National Academy of Sciences* 47, no. 10 (1961): pp. 1588–602.
30. Y. Shimizu, et al. "Cell-free translation reconstituted with purified components." *Nature Biotechnology* 19 (2001): pp. 751–55.
31. Geoff Baldwin, Travis Bayer, Robert Dickinson, Tom Ellis, Paul S. Freemont, Richard I. Kitney, Karen Polizzi, and Guy-Bart Stan. *Synthetic Biology: A Primer* (London: Imperial College Press, 2012), p. 142.
32. Mansi Srivastava, Oleg Simakov, Jarrod Chapman, Bryony Fahey, Marie E. A. Gauthier, Therese Mitros, Gemma S. Richards, Cecilia Conaco, Michael Dacre, Uffe Hellsten, Claire Larroux, Nicholas H. Putnam, Mario Stanke, Maja Adamska, Aaron Darling, Sandie M. Degnan, Todd H. Oakley, David C. Plachetzki, Yufeng Zhai, Marcin Adamski, Andrew Calcino, Scott F. Cummins, David M. Goodstein, Christina Harris, Daniel J. Jackson, et al.

"The *Amphimedon queenslandica* genome and the evolution of animal complexity." *Nature* 466 (August 5, 2010): pp. 720–26. doi:10.1038/nature09201.

33. K. W. Jeon, I. J. Lorch, and J. F. Danielli. "Reassembly of living cells from dissociated components." *Science* 167, no. 3925 (March 20, 1970): pp. 1626–27.

Chapter 10

1. Charles Darwin. *On the Origin of Species* (1859).
2. http://www.vph-noe.eu.
3. D. Noble. "Cardiac action and pacemaker potentials based on the Hodgkin-Huxley equations." *Nature* 188 (November 5, 1960): pp. 495–97.
4. D. Noble. "From the Hodgkin-Huxley axon to the virtual heart." *Journal of Physiology* 580, no. 1 (April 1, 2007): pp. 15–22. doi:10.1113/jphysiol.2006.119370.
5. See www.humanbrainproject.eu/.
6. See http://europa.eu/rapid/press-release_IP-13-54_en.htm.
7. Nobuyoshi Ishii, Kenji Nakahigashi, Tomoya Baba, Martin Robert, Tomoyoshi Soga, Akio Kanai, Takashi Hirasawa, Miki Naba, Kenta Hirai, Aminul Hoque, Pei Yee Ho, Yuji Kakazu, Kaori Sugawara, Saori Igarashi, Satoshi Harada, Takeshi Masuda, Naoyuki Sugiyama, Takashi Togashi, Miki Hasegawa, Yuki Takai, Katsuyuki Yugi, Kazuharu Arakawa, Nayuta Iwata, Yoshihiro Toya, Yoichi Nakayama, Takaaki Nishioka, Kazuyuki Shimizu, Hirotada Mori, and Masaru Tomita. "Multiple high-throughput analyses monitor the response of *E. coli* to perturbations." *Science* 316, no. 5824 (April 27, 2007): p. 593.
8. See http://wholecell.stanford.edu/.
9. See http://wholecellkb.stanford.edu/.
10. W. C. Nierman, T. V. Feldblyum, M. T. Laub, I. T. Paulsen, K. E. Nelson, J. A. Eisen, J. F. Heidelberg, M. R. Alley, N. Ohta, J. R. Maddock, I. Potocka, W. C. Nelson, A. Newton, C. Stephens, N. D. Phadke, B. Ely, R. T. DeBoy, R. J. Dodson, A. S. Durkin, M. L. Gwinn, D. H. Haft, J. F. Kolonay, J. Smit, M. B. Craven, H. Khouri, J. Shetty, K. Berry, T. Utterback, K. Tran, A. Wolf, J. Vamathevan, M. Ermolaeva, O. White, S. L. Salzberg, J. C. Venter, L. Shapiro, and C. M. Fraser. "Complete genome sequence of *Caulobacter crescentus*." *Proceedings of the National Academy of Sciences* 98, no. 7 (March 27, 2001): pp. 4136–41; see also erratum in *Proceedings of the National Academy of Sciences* 98, no. 11 (May 22, 2001): p. 6533.
11. Beat Christen, Eduardo Abeliuk, John M. Collier, Virginia S. Kalogeraki, Ben Passarelli, John A. Coller, Michael J. Fero, Harley H. McAdams, and Lucy Shapiro. "The essential genome of a bacterium." *Molecular Systems Biology* 7 (2011): article number 528.
12. Baldwin, et. al, *Synthetic Biology*.

13. T. S. Gardner, C. R. Cantor, and J. J. Collins. "Construction of a genetic toggle switch in *Escherichia coli.*" *Nature* 403, no. 6767 (January 20, 2000): pp. 339–42.

14. J. J. Tabor, H. Salis, Z. B. Simpson, A. A. Chevalier, A. Levskaya, E. M.Marcotte, C. A. Voigt, and A. D. Ellington. "A synthetic genetic edge detection program." *Cell* 137, no. 7 (2009): pp. 1272–81.

15. Tal Danino, Octavio Mondragón-Palomino, Lev Tsimring, and Jeff Hasty. "A synchronized quorum of genetic clocks." *Nature* 463 (January 21, 2010): pp. 326–30.

16. See www.clothocad.org.

17. Baldwin, et. al, *Synthetic Biology*, p. 121.

18. Karmella A. Haynes, Marian L. Broderick, Adam D. Brown, Trevor L. Butner, James O. Dickson, W. Lance Harden, Lane H. Heard, Eric L. Jessen, Kelly J. Malloy, Brad J. Ogden, Sabriya Rosemond, Samantha Simpson, Erin Zwack, A. Malcolm Campbell, Todd T. Eckdahl, Laurie J. Heyer, and Jeffrey L. Poet. "Engineering bacteria to solve the Burnt Pancake Problem." *Journal of Biological Engineering* 2, no. 8 (2008). doi:10.1186/1754-1611-2-8.

19. Parasight, Imperial College London. http://2010.igem.org/Team:Imperial _College_London.

20. Baldwin, et. al, *Synthetic Biology*, p. 121.

21. Laura Adam, Michael Kozar, Gaelle Letort, Olivier Mirat, Arunima Srivastava, Tyler Stewart, Mandy L Wilson, and Jean Peccoud. "Strengths and limitations of the federal guidance on synthetic DNA." *Nature Biotechnology* 29 (2010): pp. 208–210. doi:10.1038/nbt.1802.

22. Yael Heyman, Amnon Buxboim, Sharon G. Wolf, Shirley S. Daube, and Roy H. Bar-Ziv. "Cell-free protein synthesis and assembly on a biochip." *Nature Nanotechnology* 7 (2012): pp. 374–378. doi:10.1038/nnano.2012.65.

23. Taiji Okano, Tomoaki Matsuura, Yasuaki Kazuta, Hiroaki Suzukiac, and Yomo Tetsuya. "Cell-free protein synthesis from a single copy of DNA in a glass microchamber." *Lab Chip* 12, no. 15 (2012): pp. 2704–2711. doi:10.1039/ C2LC40098G.

24. V. Noireaux, R. Bar-Ziv, and A. Libchaber. "Principles of cell-free genetic circuit assembly." *Proceedings of the National Academy of Sciences* 100 (2003): pp. 12672–77.

25. See http://library.cshl.edu/oralhistory/interview/cshl/research/zing-finger -proteins-discovery-and-application/.

26. J. Miller, A. D. McLachlan, and A. Klug. "Repetitive zinc-binding domains in the protein transcription factor IIIA from *Xenopus oocytes.*" *EMBO Journal* 4, no. 6 (June 1985): pp. 1609–14.

27. Ahmad S. Khalil, Timothy K. Lu, Caleb J. Bashor, Cherie L. Ramirez, Nora C. Pyenson, J. Keith Joung, and James J. Collins. "A synthetic biology framework for programming eukaryotic transcription functions." *Cell* 150, no. 3 (August 3, 2012): pp. 647–58.

28. Ibid.

29. *Synthetic Genomics: Options for Governance* accessible online at www .synbiosafe.eu/uploads///pdf/Synthetic%20Genomics%20Options%20for%20 Governance.pdf.

30. See "Playing democs games to explore synthetic biology," Edinethics, at www .edinethics.co.uk/synbio/synbio%20democs%20report.pdf; and Nuffield Council on Bioethics at www.nuffieldbioethics.org/emerging-biotechnologies.

31. "Bridging science and security for biological research: A discussion about dual use review and oversight at research institutions. Report of a meeting September 13–14, 2012." American Association for the Advancement of Science in conjunction with the Association of American Universities, Association of Public and Land-Grant Universities, and the Federal Bureau of Investigation.

32. See www.biofab.org/.

33. Marcus Wohlsen. *Biopunk: DIY Scientists Hack the Software of Life* (New York: Current, 2011), pp. 65, 155.

34. Freeman Dyson. "Our Biotech Future." *New York Review of Books*, July 19, 2007.

35. See A. S. Khan. "Public health preparedness and response in the USA since 9/11: A national health security imperative." *Lancet* 378 (2011): pp. 953–56; and the "Bridging science and security for biological research" report.

36. *Biotechnology Research in an Age of Terrorism: Confronting the 'Dual Use' Dilemma* (Washington, D.C.: National Academies Press, 2004).

37. Wohlsen, *Biopunk*.

38. Baldwin, et. al, *Synthetic Biology*, p. 139.

39. There are many formulations. See Kenneth R. Foster, Paolo Vecchia, and Michael H. Repacholi. "Science and the precautionary principle." *Science* 288, no. 5468 (2000): pp. 979–81.

40. www.bioethics.gov/sites/default/files/news/PCSBI-Synthetic-Biology -Report-12.16.10.pdf.

41. Isaac Asimov. "Introduction." In *The Rest of the Robots* (New York: Doubleday, 1964).

Chapter 11

1. Arthur Conan Doyle. "The Disintegration Machine" (1929). Text accessible online at http://gutenberg.net.au/ebooks06/0601391h.html.

2. Captain Kirk in "The Gamesters of Triskelion," *Star Trek*, January 5, 1968. (Though it has become linked with the series and movies, the phrase "Beam me up, Scotty" was never spoken in any *Star Trek* television episode or film.)

3. Isaac Asimov. "It's Such a Beautiful Day." *Star Science Fiction Stories* 3 (1954).

4. C. H. Bennett, G. Brassard, C. Crépeau, R. Jozsa, A. Peres, and W. K. Wootters. "Teleporting an unknown quantum state via dual classical and Einstein-Podolsky-Rosen channels." *Physical Review Letters* 70 (1993): pp. 1895–99.

5. Xiao-Song Ma, Thomas Herbst, Thomas Scheidl, Daqing Wang, Sebastian Kropatschek, William Naylor, Bernhard Wittmann, Alexandra Mech, Johannes Kofler, Elena Anisimova, Vadim Makarov, Thomas Jennewein, Rupert Ursin, and Anton Zeilinger. "Quantum teleportation over 143 kilometres using active feed-forward." *Nature* 489 (2012): pp. 269–73. doi:10.1038/nature11472.

6. David D. Awschalom, Lee C. Bassett, Andrew S. Dzurak, Evelyn L. Hu, and Jason R. Petta. "Quantum spintronics: Engineering and manipulating atom-like spins in semiconductors." *Science* 339, no. 6124 (2013): pp. 1174–79. doi:10.1126/science.1231364.

7. A. Furusawa, J. L. Sørensen, S. L. Braunstein, C. A. Fuchs, H. J. Kimble, and E. S. Polzik. "Unconditional quantum teleportation." *Science* 282, no. 5389 (October 23, 1998): pp. 706–9. doi:10.1126/science.282.5389.706.

8. Xiao-Hui Bao, Xiao-Fan Xu, Che-Ming Li, Zhen-Sheng Yuan, Chao-Yang Lu, and Jian-Wei Pan. "Quantum teleportation between remote atomic-ensemble quantum memories." *Proceedings of the National Academy of Sciences* 109, no. 50 (December 11, 2012). doi:10.1073/pnas.1207329109.

9. J. R. Minkel. "Beam Me Up, Scotty?" *Scientific American*, February 14, 2008. www.scientificamerican.com/article.cfm?id=why-teleporting-is-nothing -like-star-trek&page=2.

10. Dimitar Sasselov. Personal communication, August 6, 2012.

11. Charles H. Townes. "The First Laser." In Laura Garwin and Tim Lincoln, eds. *A Century of Nature: Twenty-One Discoveries That Changed Science and the World* (Chicago: University of Chicago Press, 2003), pp. 107–12.

12. E. Chain, H. W. Florey, A. D. Gardner, N. G. Heatley, B. M. Jennings, J. Orr-Ewing, and A. G. Sanders. "Penicillin as a chemotherapeutic agent." *Lancet* 236, no. 6104 (August 24, 1940): pp. 226–28.

13. Edward Topp, Ralph Chapman, Marion Devers-Lamrani, Alain Hartmann, Romain Marti, Fabrice Martin-Laurent, Lyne Sabourin, Andrew Scott, and Mark Sumarah. "Accelerated biodegradation of veterinary antibiotics

in agricultural soil following long-term exposure, and isolation of a sulfamethazine-degrading *Microbacterium* sp." *Journal of Environmental Quality* 42, no. 1 (December 6, 2012): pp. 173–78. doi:10.2134/jeq2012.0162.

14. A. Fajardo, N. Martínez-Martín, M. Mercadillo, J. C. Galán, B. Ghysels, et al. "The neglected intrinsic resistome of bacterial pathogens." *PLOS ONE* 3, no. 2 (2008): e1619.

15. Gerard D. Wright. "The antibiotic resistome: The nexus of chemical and genetic diversity." *Nature Reviews: Microbiology* 5 (March 2007): pp. 175–86.

16. J. Lederberg and E. L. Tatum. "Gene recombination in *E. coli.*" *Nature* 158, no. 4016 (1946): p. 558.

17. Otto Carrs. "Meeting the challenge of antibiotic resistance." *British Medical Journal* 337 (2008): a1438.

18. Jean Marx. "New bacterial defense against phage invaders identified." *Science* 315, no. 5819 (March 23, 2007): pp. 1650–51.

19. Cairns, Stent, and Watson, *Phage and the Origins of Molecular Biology*, p. 5.

20. See http://newsarchive.asm.org/sep01/animalcule.asp.

21. See Alexander Sulakvelidze, Zemphira Alavidze, and J. Glenn Morris, Jr. "Bacteriophage therapy." *Antimicrobial Agents Chemotherapy* 45, no. 3 (March 2001): pp. 649–59. doi:10.1128/AAC.45.3.649-659.2001.

22. F. d'Herelle. "Sur un microbe invisible antagoniste des bacillus dysentérique." *Comptes rendus de l'Académie des Sciences* 165 (1917): pp. 373–75.

23. E. H. Hankin. "L'action bactericide des eaux de la Jumna et du Gange sur le vibrion du cholera." *Annales de l'Institut Pasteur* 10 (1896): pp. 511–23.

24. See Eliava Institute at www.eliava-institute.org/?rid=2.

25. Cairns, Stent, and Watson, *Phage and the Origins of Molecular Biology*, p. 5.

26. Editorial, "Limitations of bacteriophage therapy." *JAMA* 96 (1931): p. 693; and editorial, "Commercial aspects of bacteriophage therapy." *JAMA* 100 (1933): pp. 1603–4.

27. William C. Summers. "The strange history of phage therapy." *Bacteriophage* 2, no. 2 (2012): pp. 130–33.

28. G. S. Stent. *Molecular Biology of Bacterial Viruses* (San Francisco and London: W. H. Freeman, 1963), pp. 8–9.

29. Summers, "The strange history of phage therapy," pp. 130–33.

30. Lauren Gravitz. "Turning a new phage." *Nature Medicine* 18 (2012): pp. 1318–20. doi:10.1038/nm0912-1318.

31. Christopher Browning, Mikhail Shneider, Valorie Bowman, David Schwarzer, and Petr Leiman. "Phage pierces the host cell membrane with the iron-loaded spike." *Structure* 20, no. 2 (February 8, 2012): pp. 326–39.

32. M. S. Zahid, S. M. Udden, A. S. Faruque, S. B. Calderwood, J. J. Mekalanos, et al. "Effect of phage on the infectivity of *Vibrio cholerae* and emergence of genetic variants." *Infection and Immunity* 76 (2008): pp. 5266–73. Jean Marx. "New bacterial defense against phage invaders identified." *Science* 315, no. 5819 (March 23, 2007): pp. 1650–51.

33. Sheena McGowan, Ashley M. Buckle, Michael S. Mitchell, James T. Hoopes, D. Travis Gallagher, Ryan D. Heselpoth, Yang Shen, Cyril F. Reboul, Ruby H. P. Law, Vincent A. Fischetti, James C. Whisstock, and Daniel C. Nelson. "X-ray crystal structure of the streptococcal specific phage lysin PlyC." *Proceedings of the National Academy of Sciences* 109, no. 31 (July 31, 2012). doi:10.1073/pnas.1208424109.

34. A. Wright, C. H. Hawkins, E. E. Anggård, and D. R. Harper. "A controlled clinical trial of a therapeutic bacteriophage preparation in chronic otitis due to antibiotic-resistant *Pseudomonas aeruginosa*: A preliminary report of efficacy." *Clinical Otolaryngology* 34, no. 4 (August 2009): pp. 349–57.

35. L. J. Marinelli, et al. "Propionibacterium acnes bacteriophages display limited genetic diversity and broad killing activity against bacterial skin isolates." *mBio* 3, no. 5 (2012): e00279-12.

Chapter 12

1. Isaac Newton. *Opticks*, 2nd edition (1718). Book 3, Query 30, 349.

2. Brett J. Gladman, Joseph A. Burns, Martin Duncan, Pascal Lee, and Harold F. Levison. "The exchange of impact ejecta between terrestrial planets." *Science* 271, no. 5254 (March 8, 1996): p. 1387(6).

3. David S. McKay, et al. "Search for past life on Mars: Possible relic biogenic activity in Martian meteorite ALH84001." *Science* 273, no. 5277 (1996): pp. 924–30.

4. I. P. Wright, M. M. Grady, and C. T. Pillinger. "Organic materials in a martian meteorite." *Nature* 340 (July 20, 1989): pp. 220–22.

5. Carnegie Institution Geophysical Laboratory Seminar, May 14, 2007. Summarized in Gilbert V. Levin. "Analysis of evidence of Mars life." *Electroneurobiología* 15, no. 2 (2007): pp. 39–47. Ronald Paepe. "The red soil on Mars as a proof for water and vegetation (PDP)." *Geophysical Research Abstracts* 9, no. 1794 (2007). R. Navarro-González, et al. "The limitations on organic detection in Mars-like soils by thermal volatilization–gas chromatography–MS and their implications for the Viking results." *Proceedings of the National Academy of Sciences* 103, no. 44 (2006): pp. 16089–94. See also: Rafael Navarro-González, Edgar Vargas, José de la Rosa, Alejandro C. Raga, and Christopher P. McKay. "Reanalysis of the Viking results suggests perchlorate and organics at midlatitudes on Mars." *Journal*

of Geophysical Research 115 (2010); accessible online at www.earth
.northwestern.edu/individ/seth/438/mckay.viking.pdf.

6. See www.nasa.gov/mission_pages/msl/multimedia/pia16576.html.

7. M. Carr and J. Head. "Oceans on Mars: An assessment of the observational
evidence and possible fate." *Journal of Geophysical Research* 108 (2003): p.
5042. doi:10.1029/2002JE001963.

8. See http://marsrover.nasa.gov/newsroom/pressreleases/20121204a.html.

9. Marc Kaufman. "Mars Curiosity Rover Finds Proof of Flowing Water—A
First." *National Geographic*, September 27, 2012. http://news
.nationalgeographic.com/news/2012/09/120927-nasa-mars-science
-laboratory-curiosity-rover-water-life-jpl/.

10. Francis M. McCubbin, Erik H. Hauri, Stephen M. Elardo, Kathleen E. Vander
Kaaden, Jianhua Wang, and Charles K. Shearer Jr. "Hydrous melting of the
Martian mantle produced both depleted and enriched shergottites." *Geology*
G33242.1 (June 15, 2012).

11. Conference on the Geophysical Detection of Subsurface Water on Mars (2001).

12. M. Max and S. M. Clifford. "Mars methane: A critical in-situ resource to
support human exploration." Concepts and Approaches for Mars Exploration
(2012). www.lpi.usra.edu/meetings/marsconcepts2012/pdf/4385.pdf.

13. Morten E. Allentoft, Matthew Collins, David Harker, James Haile, Charlotte
L. Oskam, Marie L. Hale, Paula F. Campos, Jose A. Samaniego, M. Thomas P.
Gilbert, Eske Willerslev, Guojie Zhang, R. Paul Scofield, Richard N. Holdaway,
and Michael Bunce. "The half-life of DNA in bone: Measuring decay kinetics
in 158 dated fossils." *Proceedings of the Royal Society B* 279, no. 1748
(December 7, 2012): pp. 4724–33.

14. Yudhijit Bhattacharjee. "Failure to Launch: Mars Missions Sidelined in New
NASA Budget Proposal." *Science*, Science Insider blog, February 13, 2012.
http://news.sciencemag.org/scienceinsider/2012/02/failure-to-launch-mars
-missions.html?ref=hp.

15. The report of the Commission of Inquiry on the loss of the *Beagle 2* mission is
accessible at www.bis.gov.uk/assets/ukspaceagency/docs/space-science/
beagle-2-commission-of-inquiry-report.pdf.

16. H. Price, et al. "Mars Sample Return Spacecraft Systems Architecture." Jet
Propulsion Laboratory. California Institute of Technology. http://trs-new.jpl
.nasa.gov/dspace/bitstream/2014/13724/1/00-0092.pdf.

17. The UN treaty is accessible online at www.oosa.unvienna.org/oosa/SpaceLaw/
outerspt.html.

18. Michael Crichton. *The Andromeda Strain* (New York: Knopf, 1969).

Index

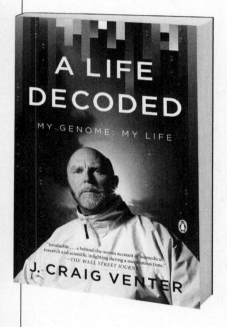